MESSAGE IN A BOTTLE

Ocean Dispatches from a Seabird Biologist

HOLLY HOGAN

ALFRED A. KNOPF CANADA

PUBLISHED BY ALFRED A. KNOPF CANADA

Copyright © 2023 Holly Hogan
All illustrations © Ian L. Jones: p. xx The Extinct Great Auk; p. xx Turtle
Pursuing a Balloon; p.xx Drowning Murre in Gillnet; p. xx Seabird in Sargassum;
p. xx Abandoned Whaling Station at Stromness

www.penguinrandomhouse.ca

LIBRARY AND ARCHIVES CANADA CATALOGUING IN PUBLICATION
Title: Message in a bottle : ocean dispatches from a seabird biologist / Holly Hogan.
Names: Hogan, Holly (Biologist), author.
Identifiers: Canadiana (print) 20220430284 | Canadiana (ebook) 20220430292 |
    ISBN 9780385696265 (hardcover) | ISBN 9780385696272 (EPUB)
Subjects: LCSH: Plastic marine debris. | LCSH: Plastic marine debris—
    Environmental aspects. | LCSH: Marine pollution. | LCSH: Marine pollution—
    Prevention. | LCSH: Marine ecology.
Classification: LCC TD427.P62 H64 2023 | DDC 363.739/42—dc23

Book design: Matthew Flute
Image credits: Westend61/ Getty Images

Printed in Canada

10 9 8 7 6 5 4 3 2 1

Penguin
Random House
KNOPF CANADA

For Arielle, Robin and Ben

MESSAGE IN A BOTTLE

People protect what they love.

—JACQUES-YVES COUSTEAU

. . . when the last individual of a race of living things breathes no more, another heaven and another earth must pass before such a one can be again.

—WILLIAM BEEBE

MAIN OCEAN SURFACE CURRENTS

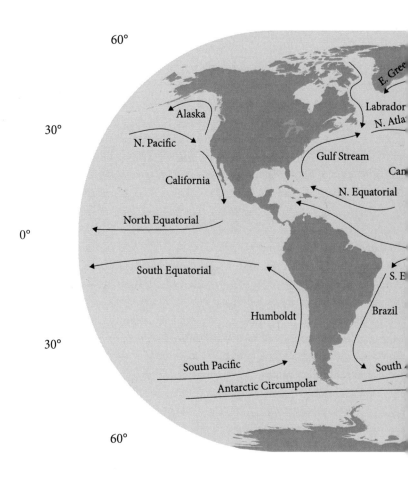

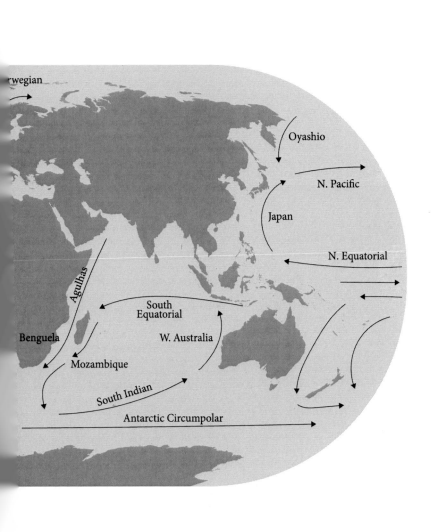

Contents

Introduction: Apex Predator   x

Part 1: The Ocean's Secrets   x

     1. Throwaway Living   x

     2. Wilderness, Rewritten   x

     3. The Heart of the Labrador Current   x

Part 2: The Fragility of Marine Life   x

     4. Quiet at Witless Bay   x

     5. Full Throttle on Baccalieu Island   x

     6. Ghost Harvest: The Threat of Macroplastics   x

     7. Right Whale, Wrong Place: The Plight of
         the North Atlantic Right Whale   x

Part 3: Ocean Currents and the Plastic Crisis   x

     8. Kittiwakes and the Japan Current   x

     9. Following Currents to the Ends of the Earth   x

     10. The Raven's Parachute: Plastic and the Gulf Stream   x

     11. The Plastisphere: Microplastics,
         the Oceans and Human Health   x

Part 4: Petrochemicals and the Future of Our Seas   x

     12. Whales on Toast   x

     13. Out of Sight, Out of Mind   x

     14. Reframing Plastic: The Potential of a Circular Economy   x

     15. New Beginnings: Land of Ice and Hope   x

     16. Fabric from Fairy Rings   x

Sources   x

Acknowledgements   x

Index   x

Introduction:

Apex Predator

The Zodiac driver gunned the engine, working with the incoming waves to plough the boat as far up onto the beach as possible. The water was frigid, the wind a strong northeasterly, pounding the waves against the shore as the metal floor of the inflatable boat scraped and rumbled to an abrupt halt on the beach. We had just travelled across a wide expanse of open water, squinting against the glare of the sun. Water splashed our bare hands, which were frozen into fists around the safety rope that ran along the gunnel—protection against being bounced out of the boat. We gathered our gear and piled out over the hull, one at a time with practised efficiency, before the suck of the withdrawing waves could pull the boat into deeper water and over our boots, which would make for pretty cold and uncomfortable feet at best. Complete submersion—well, you just wanted to avoid that altogether. We unloaded the landing gear and prepared to do a final safety check before bringing guests ashore.

That's how I arrived with an expedition team on Killiniq Island at the northern tip of Labrador in the fall of 2016—a pretty typical landing for ship-based travel in wild places. Extreme conditions can make polar regions more challenging than most. At the exposed tip of Labrador, the wind is a given, the sunshine a gift. This was a singular day.

We gathered above the landwash and the expedition leader assigned responsibilities among the team. I grabbed my telescope and backpack and headed toward the backshore where a trail led to an abandoned radar station and its attendant buildings in various states of decay—a point of interest for the history buffs. In my role as wildlife biologist, I would be looking for any natural features or phenomena to train my scope on. Flocks of sea ducks, maybe some seals. Or maybe something interesting on the ground that might tell a story: Owl pellets packed tight with hair and bones. A lemming trail snaking through the compact vegetation, disappearing into a burrow. That kind of thing. But what I saw stopped me in my tracks. It was not that kind of thing.

We had started the journey to Labrador in Kangerlussuaq on the west coast of central Greenland, the point of departure for many Arctic expedition trips. I was working for Adventure Canada, a small family-run expedition company that offers adventure travel to the Arctic and beyond, hired for my seabird experience. On a charter flight from Toronto to Kangerlussuaq, we flew over the edge of Greenland's polar ice cap, a sheet of glacier-covered mountains as far as the eye could see. It is one of only two ice sheets on the planet, covering 1.7 million square kilometres, almost the entirety of the island. Our planet depends on the ice sheets to control climate, maintain sea level, nourish global marine ecosystems and provide over half the air we breathe; they are critical to life on earth as we know it. From the air, it looked like this one was up to the challenge. Impenetrable, indifferent. Inviolate. It was hard to conceive that this endless expanse of ice that had existed for over 100,000 years was now starting to disintegrate.

The plane began its descent. It was September and the Arctic willow was turning from green to gold, the ground birch a vibrant red—the landscape on fire. The head of the fjord was an estuary braided with streams of pale, murky glacial silt. Fresh snow had gathered on the crags and ledges of the surrounding mountains, like marbled fat. Kangerlussuaq is at the end of the spectacular ninety-kilometre fjord that grazes the Arctic Circle.

It was dark by the time we gathered on the dock. The frenetic pace of starting a new journey greeted us there—expedition team staff with headlamps, searching for the joining team members, handing off duties. *Hi, Holly, right? Take my headlamp—you are to help guests board the tender boats—ten to a boat. Make sure they have their personal bags with them and their life vests are done up properly. Good luck!* After a brief and hectic sojourn on land, we boarded the *Sea Adventurer* and sailed for hours through the stillness of the fjord and into the night.

It takes the better part of two days to cross the North Atlantic to Labrador at this latitude, an expanse of ocean known as the Davis Strait. The frenetic midsummer hubbub of marine activity that comes with feeding frenzies during the warmer months had passed. And besides, the deep open ocean is less active than most people expect. There are amazing experiences to be had, but you have to work for them.

Early in the crossing I gave a presentation on the dovekie, a stocky little seabird that is closely related to the puffin. Black on the back, white on the belly, the size of a child's heart. What they lack in size they more than make up for in toughness, thriving in one of the harshest environments on earth. Most spend their winters off the coast of Newfoundland, facing head-on whatever the stormy North Atlantic has to throw at them, sheets of blowing spray and crushing walls of six-metre waves a regular occurrence. The majority of the world's population breed by the tens of millions in the scree slopes of the Thule district of northwest Greenland, taking advantage of the relative proximity to the Pikialasorsuaq—meaning "great upwelling" in Greenlandic. This great upwelling creates an area of open water year-round, better known as a polynya. In English, this particular one is known as the North Water Polynya. It is the polar region version of a desert oasis, a rich and essential food source for many marine animals—predator and prey. For dovekies, it is an all-you-can-eat buffet of an Arctic copepod, the small but super-fat sea creature that they feast on. And once their egg hatches, the parents take turns stogging their throat pouch with extras to feed their fast-growing and demanding chick.

At the end of the breeding season, the parents slack off on the chick feedings. In a dovekie version of "here's your hat, what's your hurry?" the parents are sending the signal that it's time for them to start fending for themselves. When the chick decides to take on whatever lies beyond the safety of its stony burrow, one parent guides it out to sea and over the next few weeks teaches it how to navigate the North Atlantic's tempestuous storms and ephemeral distribution of tiny prey. They use the Labrador Current to advantage, hitching a ride to the west Atlantic, down the Labrador coast and onward to the east coast of Newfoundland. They are designed for diving; flight is an afterthought.

Even after years of encountering them at sea, I am still moved by the dovekies' diminutive size in stark contrast to their enormous grit and determination. No matter what my mood, the sight of a dovekie makes me smile from the inside out.

With millions on the move, the chances of seeing dovekies were pretty good and I led the guests out on deck to try. It is a great bit of sport. It takes a while to spot the first one; no one expects something that tough to be so impossibly small. Some are quicker to get the knack. The newly experienced spotters helped the others, the bow crowded with dovekie enthusiasts, the roar of the wind regularly punctuated with "There's one!" or "Two together!" By September, the adults and chicks are about the same size—distinguishable mostly by their behaviour, the nervous parent gesturing to the chick and finally diving out of the way of the ship. The inexperienced chick hesitates for a beat ("oh!") before following its parent's example.

The whole group is laser-focused on the ocean surface, searching with great purpose and delight for something they hadn't heard of more than a couple of hours ago.

Aside from the predictable sights on a sea voyage, there is what is left to chance: a whale blow, or maybe a peregrine falcon, trading one side of the Atlantic for the other. Still, no matter how much enthusiasm you manage to squeeze out of a long stretch at sea, everyone is excited by the sight of land and a chance to explore off the ship.

The coast of the Canadian Arctic differs from most of Greenland in one important way: the chance of encountering polar bears. During the crossing, guests and staff alike were required to attend a polar bear safety briefing. Before landing, the site was scanned from the ship to determine if a polar bear could be seen nearby. If the coast was clear, armed bear monitors went ashore to "sweep" the area, ensuring none were hidden behind a landscape feature or an abandoned building. Only then did the rest of the expedition team come ashore.

When we arrived at Killiniq Island, a bear had been seen on another small island nearby, a distance deemed non-threatening. The bear monitors spread around the periphery of the landing, defining the boundary beyond which no one was allowed to venture. Once the guards were in place, I headed up the beach with the rest of the expedition team to explore for a position of best advantage.

The thing that had stopped me in my tracks as I walked up from the landing site was this: a large, stinking mound of fresh polar bear dung. At its centre, a completely intact disposable baby diaper. My first thought: *What happened to the baby?*

It quickly became obvious that there had been no human tragedy here. The tragedy was the diaper itself: the gleaming white plastic, its waistband decorated with pink, puffy clouds, encasing a space-age, super-absorbent plastic material. Plastics wrapped in plastic.

Used diapers are essentially a soiled mess of plastic compounds that cannot be easily recycled, and over four million tons of them end up in landfills or are incinerated every year. But where did this diaper come from? The nearest community was Iqaluit, on Baffin Island—about 400 kilometres to the north, on the other side of the Hudson Strait. It had probably bobbed over from Iqaluit, or somewhere else on Baffin Island, carried by the same currents that guided the dovekies.

Polar bears are considered marine mammals because of their dependence on hunting from sea ice for survival. In the fall of the year, after the ice-free summer months, polar bears are at their leanest and closest to starvation. They move slowly, if at all, to conserve energy until the ice forms and they are able to hunt seals again. They scavenge what they can and will even eat seaweed to fill their stomachs. This bear had been duped by the promising smell of something organic. Instead of protein, it had devoured an expanded ball of plastic. Luckily, the diaper remained intact, maybe offering a few hours of false relief before passing through the system entirely.

Polar bears are apex predators, at the top of the marine food chain. They specialize on seals and are designed for the feast-or-famine availability of them. Still, they are able to eat pretty much anything on offer. They are equipped to endure whatever punishment the Arctic environment dishes out, but I can't help feeling that the diaper has outmatched the polar bear for its ability to persevere.

Plastic has found its way into the most remote regions of the planet, travelling on ocean currents and falling like snow to the ocean's depths. Natural forces are not driven by the vagaries of human will or desire but are profoundly affected by them. It has taken me the better part of a lifetime to appreciate not only the sublime beauty and richness of the ocean but the unintentional impact our choices are having. It is only by sailing through remote areas and seeing some of that impact for myself that I have come to appreciate the immense scale of the problem. And it wasn't the diaper that did it. It would be another year before I felt the devastation personally and profoundly. That the abundance is not infinite, and that human activities are pushing the marine ecosystems we rely on to the brink of collapse.

HAPPY
VALENTINE'S DAY
CARETTA
YOURS FOREVER

PART

One

The Ocean's Secrets

Chapter 1

Throwaway Living

The ocean is slow to reveal its secrets. Its endless horizons and seemingly bottomless depths suggest an infinite capacity to hold. And to withhold. Plastic has been one of its well-kept secrets, mostly hidden below the surface and out of sight. I have spent much of my life living and working in Newfoundland, a remote place with half a million people and 12 million seabirds, give or take. Outnumbered 24:1, it was easy to take the natural abundance for granted. In more recent years I have sailed to the Arctic, the Antarctic and every latitude in between. I've been carried and pulled off course by the currents that wrap around the globe and observed the influence that they have. The oxygen and nutrients they circulate, the life they support—each current individual and unique, but also completely dependent on and interconnected with all the others.

I started to conduct ship-based seabird surveys regularly in 2014 for the Canadian Wildlife Service, an agency of the Department of

Environment and Climate Change Canada. The surveys are part of a larger program that monitors seabird populations and distribution over time. When the surveys are taken together, over all seasons and many years, patterns of distribution emerge, showing where seabirds are found at different times of the year and any changes in these patterns that may occur, particularly in response to climate change. Where possible, behavioural data are also collected—feeding, caring for young, resting, migrating—to reveal not only where the birds are but why they are there. Seabirds at the surface can be directly linked to the ocean temperature, chemistry and marine life below the surface; they are a handy indicator of what might be going on in the ocean's depths.

I was spending several months a year at sea aboard Canadian, American and European research vessels. Much of the time was spent far offshore, primarily to account for seabird distributions at sea throughout the year, but I also recorded everything else that came in the ship's path, including whales, seals, turtles—and garbage. I was surprised by the floating plastic debris I saw. There were days when I recorded more garbage than seabirds, hundreds of kilometres off the coast. Sometimes they were together—an Arctic tern sitting on a lost buoy, a fulmar picking at bright-pink plastic in a mat of floating kelp. Still, it was not the quantity that alarmed me so much as the unlikely locations. What is a plastic Javex bottle doing in the middle of the Atlantic Ocean? A shadow of apprehension was slowly engulfing me.

On Valentine's Day in 2017 I was doing seabird surveys on an American National Oceanographic and Atmospheric Administration (NOAA) ship off the coast of New York City. Loving tributes were scattered across the water: *Love You Forever. Be Mine! With My*

Whole Heart. Messages on Mylar balloons with elaborate ribbon trains, scattered with the wind and arriving, deflated, on the ocean surface. Near one of these balloons, a subtle disturbance caught my eye. It was a loggerhead turtle, making slow but determined progress toward the balloon.

Turtles feed primarily on jellyfish. A deflated balloon with a ribbon tail is a dead ringer for a jellyfish. There were three possible outcomes here: The turtle could be outpaced by the wind and surface current that was carrying the balloon along the surface. It could catch up with the balloon, eat it and subsequently choke. Or it could get tangled in the ribbons and drown. The ship kept steaming on its course and I don't know what happened in the end. But my encounter with this turtle and the plastic buffet surrounding it deepened my understanding of just how lethal our carelessly discarded plastic can be to the marine life that encounters it. The diaper that had been eaten by a polar bear did not affect me in the same way, perhaps because the bear appeared to be unscathed by the encounter. The idea that plasticizers and other plastic toxins might have leached into its bloodstream was an unknown threat to me then. My awareness started at the ocean's surface, before going deeper.

By 2017, I had spent a good deal of time at sea. Still, my experiences to that point were mostly in the Labrador Current and Arctic Ocean—marine waters that did not immediately suggest the global crisis, largely because of the small populations that inhabit the adjacent coastlines and the nature of the currents that move water bodies around that region. But it had become clear that plastic was finding its way to the most remote regions on earth. The world's oceans—and the human garbage we have dumped in them—are all interconnected by the currents that move water around the globe.

There is really only one ocean, and what happens to it in one region affects all others.

All the marine life I was seeing—seabirds, whales, seals and turtles—is designed for a life spent at sea. Even for these animals, the ocean is a challenging environment to survive in. Some of the threats arise from the naturally occurring vagaries of life in a harsh and unpredictable environment, including changes in food availability from one year to the next, poor weather conditions at a critical time—particularly during the breeding season—and predation. Young, inexperienced animals are at greater risk of all sources of natural mortality, but if they make it out of this naive and fraught period, most are long-lived. There is lots of variation, but for the most part seabirds live at least twenty years, with a breeding window of about fifteen years. For a population to remain stable, two adults just need to replace themselves over their lifetime. And though the odds are stacked against survival of the young, the longevity of the adults means that it generally works out.

But if you overlay these naturally occurring challenges with manmade ones, marine life can get into serious trouble. Oil spills. Collision with ships and oil platforms. Climate change's effect on ocean temperatures and the resulting change in prey availability and distribution. Entanglement in fishing gear. Entanglement in plastic garbage. Marine toxins. Ingestion of plastics. Seabirds face additional challenges. On breeding colonies, you can add habitat degradation and destruction. Even light pollution can pose problems. Some seabirds are attracted to light for various adaptive reasons, and artificial light, both on land and at sea, can lead them dangerously astray. Some of these challenges have been around for decades; some are just emerging. And each is exerting pressures in different ways.

Plastic has been around for decades, but the explosion in its use, particularly its single use, has brought about a modern crisis.

Until I immersed myself in research, I knew very little about the devastating impact the accumulating plastic from our day-to-day lives was having on marine animals. I had seen marine wildlife in close association with marine plastic many times. But it wasn't until 2017, when I saw the balloon-as-love-letter-with-whimsical-ribbon in a potentially lethal collision with a hungry turtle, that I was startled into awareness of the much larger issue at play.

———

The word *plastic* came from the ancient Greek *plassein*, "to mould" and it can be used to describe anything that is malleable and easily shaped, from ideas to objects. Plastic has, for the most part, been co-opted to define the synthetic materials we know as plastic today. And appropriately so—plastic can be used to make just about anything.

The earliest form of plastic was created by the American inventor John Wesley Hyatt in 1869, in response to a competition to come up with a substance that would replace ivory for the manufacturing of billiard balls. The game had gained popularity at the time, but ivory required the slaughter of elephants for their tusks, forcing their populations to dangerously low levels. As a result, ivory had become expensive and increasingly hard to come by. To the relief of the middle class and elephants everywhere, Hyatt succeeded in developing a replacement product. Using a combination of cellulose from plant fibres and nitrogen from camphor, he created a new substance called celluloid. Celluloid is regarded as the first-ever plastic, but it was made from naturally occurring materials and had limited application. It also had the unsettling quality of occasionally exploding on impact. Not ideal, really, for a game designed around colliding balls.

The major innovation leading to modern plastics was made in 1907 by a Belgian chemist, Leo Baekeland. He was experimenting to find a synthetic substitute for shellac, a sticky substance produced by the lac beetle and used to insulate electrical wires. The demand for electricity was accelerating rapidly at this time and there was a need for a replacement product that could be mass-produced to meet the growing need. Baekeland combined formaldehyde and phenol, a waste product of coal, and subjected the mixture to heat and pressure. The resulting compound was a completely synthetic substance, not found anywhere in nature, which became known as Bakelite. Heat-resistant and durable, Bakelite could be mass-produced and moulded to almost any shape.

Hyatt and Baekeland had each been inspired to create a new material to solve a specific problem, but the potential for new materials and markets had the chemical industry spinning. The race was on to create as many plastic materials as possible, trusting that uses for them could be found later.

They found them.

Modern plastics are petroleum-based products. Petroleum and other fossil fuels are often referred to as hydrocarbons, simply because they are made up of hydrogen and carbon atoms. Through chemical processes, plastics are made by rearranging these elements into different shapes, forming unique molecules, called monomers. Millions of these monomers join, forming long strands that are bonded together to form polymers. Other chemicals and additives can then be incorporated into the plastic to create a desired quality and texture. The names of different plastics reflect the repeating chain structure, starting with *poly*, from the Greek word for many: polyester, polyethylene, polystyrene—you get the idea. The rest of the name is shorthand for the unique molecular structure of the monomer. If you speak "chemistry" you can figure out the structure from the name. All plastic polymers have strong carbon bond structures that make them durable and lightweight. In addition to these convenient qualities, plastics are also sterile, cheap, endlessly adaptable, and they can be produced in huge quantities to make products that are affordable for the masses. Their sterile and shatterproof properties have made plastic polymers invaluable in the scientific research and medical fields.

In August 1955, *Life* magazine published an article celebrating the dawn of a new age of disposable products called "Throwaway Living," with the subtitle "Disposable Items Cut Down Household Chores." A photo accompanying the article shows an ecstatic family of three, gathered around a trash can, throwing single-use products into the air like confetti—frozen food containers, paper napkins, disposable diapers, tinfoil pans, paper cups. The article goes on to exclaim, "The objects flying through the air in this picture would take 40 hours to clean—except that no housewife need bother."

I read that line to my husband, Michael. He asked, "Is that because the husband's offered to do it instead?"

Not likely. The article reflects the times, and so do the products in the photo, which are conveniently disposable, but mostly made from paper and tinfoil. The petrochemical industry was quick to catch up, however. The 1950s saw the beginning of the explosive demand for throwaway products, and along with that came the development and mass production of cheap disposable plastic. There was an easy answer to the question of how to deal with all the plastic: throw it in the garbage.

Plastics are now used in virtually everything, from children's toys to NASA's Mars Exploration Rovers. Single-use plastic is particularly attractive, providing huge convenience at a very low cost. Plastic straws, cutlery, takeout food containers, shopping bags, food packaging, packaging of all kinds for that matter, water bottles, tampon applicators—the list is endless. Sterile packaging has created the mentality that food is not safe to eat unless it is wrapped in plastic. We have become addicted to the stuff, numb to how pervasive it is in our daily lives. I know I certainly was. Though I was increasingly conscious of plastic in the environment, I didn't quite see my own dependency on it. At least, not until the summer of 2016.

It was early summer, the height of the forest bird breeding season, and months before I encountered the polar bear diaper excrement. I led a field team tasked with finding bird nests along the power transmission corridor of the Muskrat Falls hydroelectric megaproject development on the Churchill River in Labrador. For part of the summer, we stayed in a rented house in the small community of Sunnyside. There wasn't a local recycling facility, so I cleaned my plastic and placed it on the kitchen counter, to be taken home at the end of the week. Within a couple of days, the counter was overflowing with my single-serving yogurt containers and packaging from microwaveable meals. My favourite was the brand that came in two parts: a microwaveable plastic steamer on top of a separate plastic bowl. A healthy meal conceived and consumed within ten minutes, the plastic then washed and put aside. I was taken aback by the quantity I was producing, but not concerned. It was all recyclable (it carried the recycle symbol, after all) and it would all be taken care of.

Those reassuring chasing arrows lulled me into the false sense that my plastic would be dealt with in an environmentally responsible way. But it still has to go somewhere. The problem, I would soon understand, is that there is no "away." And in practice, most of the recyclable plastic never sees the inside of a recycling facility.

The durability of plastic, one of its greatest assets, also carries with it a serious inherent problem: plastic doesn't break down. The effects of weathering, UV radiation and other stressors simply cause the plastic polymer strands to break apart into smaller and smaller pieces, until they eventually become what are known as microplastics, which break up into even smaller nanoplastics, invisible to the naked eye. For this reason, plastic has been accumulating in the environment ever since the first plastics were made.

Imagine plastic polymers as a cake. You mix the ingredients together and bake them into the delightful confection that bears no resemblance to its original ingredients. You can break a cake into infinitely smaller crumbs, but it will never revert to its base ingredients: eggs, flour, yeast, butter. The cake remains cake, just as plastic remains plastic, no matter how small the piece.

There are now research efforts aimed at finding ways of breaking plastics down, primarily with the use of microbes—more on that later. But whether or not breaking down plastic is a good solution is complicated. Turning petroleum into plastic—through extraction, transport and production—results in enormous emissions of greenhouse gases. To put it in perspective, if plastic was a country (Plastispherica?), it would be the fifth-largest emitter of greenhouse gases in the world, exceeded only by China, the US, India and Russia. Like all petroleum products, plastic releases CO_2 when it is burned. Similarly, most of the proposed solutions involve the creation and release of more CO_2 into the atmosphere. The persistence of plastic in the environment is clearly a problem, but so is breaking it down. Maybe the biggest problem of all is that there is so much of it.

Since the 1950s, the post-Bakelite plastic frenzy has been accompanied by a global population increase from 2.5 billion to nearly 8 billion people and counting. With our increased dependence on plastic and a coinciding population explosion, we find ourselves in an environmental crisis.

The statistics regarding plastic waste are dire and disheartening. Consider plastic beverage bottles alone: Over one million of them are sold every minute. Of these, only 8 percent are recycled. The rest end up as litter or in landfills, leaching plastic toxins and particles into groundwater and streams and inevitably ending up in the ocean.

In areas of the world with poor or no waste management systems, mountains of plastic accumulate on land, overflowing into rivers and lakes, which eventually flow into the ocean. One dump truck load of plastic enters the ocean every minute, 11 million tons a year. In some countries—India, China and Indonesia at the top of the list—plastics are routinely dumped directly into the ocean. A recent study reports that about five *billion* metric tons of plastic waste have accumulated on the earth to date.

According to a report published by the European Parliament in October 2018, 82 percent of all marine litter is plastic, amounting to 123 million metric tons of plastic waste in the ocean. Of this, half is single-use plastic.

If current plastic production and waste management trends continue, the amount of plastic on the earth will more than double to

12 billion metric tons by 2050. How much of this will end up in the ocean? About 973 million tons—at which point there will be more plastic in the ocean than fish.

We have arrived at this calamity by small increments, unaware. Like the allegory of the baby crab that crawled into a bottle on the ocean floor, safe from predators and currents, and with plenty of marine debris for food. It thrived there, until one day, feeling the urge to migrate, it ventured to the mouth of the bottle—only to discover that it had grown too large to fit through. The bottle that had once offered a safe haven had become its coffin. Like the crab, we have embraced plastic's convenience and the improved lifestyle it promises, without giving a thought to any consequences it might bring.

Chapter 2

Wilderness, Rewritten

My career in the wildlife field began with a series of lies. Being the second youngest of five children in a strict Catholic family, I was good at it. I'd had lots of practice.

My father was a physician and kept magazines in his office waiting room—*Good Housekeeping, Chatelaine*, the usual fare for an office waiting room in the 1970s. When they started to pile up, he would bring them home for us. There was one odd gem in the mix, *International Wildlife*. I devoured the thing from cover to cover. First I pored over every luscious, exotic photo, and then I flipped back to the beginning and read each article in order. By grade five I was the undisputed class expert on all things furry, slimy or feathered. Orangutans, whales, emus, moose. Eventually my father got me my own subscription to the magazine.

We lived in St. John's, Newfoundland, and the wildlife I admired on the page was nowhere to be found in my immediate environment.

Like most islands, Newfoundland has less diversity than the mainland. To feed my fascination, I watched the wildlife infomercial series *Hinterland's Who's Who*, produced by the Canadian Wildlife Service. Each ad ended the same way: "For more information, why not contact the Canadian Wildlife Service, Ottawa?"

Why not indeed! I wrote to the Canadian Wildlife Service in Ottawa and they sent me a list of their *Hinterland Who's Who* pamphlets—a series dedicated to Canada's wildlife, each one describing the life history of the animal illustrated on the cover. I wanted them all, but didn't want to be greedy. What did I want most? It was sweet torture. I finally settled on twelve out of the list of about thirty. Cougar, robin and wolf among them. When the order arrived, I read, numbered and filed them on my bookshelf.

In 1982, my first year of university, I was eligible to apply for a summer position as a naturalist at the Salmonier Nature Park, which features large natural enclosures for wildlife found in Newfoundland. At seventeen, I was well-read and knew more than most kids my age; however, I had never worked anywhere before and had no relevant experience. But this was my dream job. Clearly, some embellishment was required.

On the application, you had to rate yourself on a scale of one to five on a series of skill sets. If you awarded yourself a high ranking, you had to support the claim. I was no fool—I knew average or below would get me nowhere. So I excelled at pretty much everything. For example:

Writing: 5/5. I had won a story-writing contest. (Well, not exactly. In fact, not at all.)

Artistic skills: 5/5. I had won a wildlife conservation poster contest. (That was actually true. It featured a quote about extinction from the American naturalist William Beebe, surrounded by birds I drew from photos I had found in the *Encyclopaedia Britannica*.)

First Aid: 2/5. A little more generous than I deserved, but I wasn't going to lie about safety. I was not completely lacking in moral fortitude.

Bird identification: 4/5. I'm a birdwatcher. (This was also a complete fabrication. But really—how many birds could there be? If I got an interview, I'd just learn them.)

I got an interview—better learn the birds. There was going to be a quiz, so I took Peters and Burleigh's *Birds of Newfoundland* out of the library. Two hundred and twenty-seven species? Surely not. I learned as many as I could and finally decided that if shown a yellow bird, I would just call it a warbler. Not a bad strategy given the odds, but not the stuff of an actual birdwatcher.

The first question of the first job interview of my life: "It says here you won a story-writing contest. Could you tell us about that story?"

I had worn a warm sweater and quickly realized that had been a mistake. "Oh, that old thing. Well ..." And I proceeded to tell a story that I realized, soon after the heat was off, bore a striking resemblance to *Bambi*. The bird quiz was only two slides, a robin and a great horned owl.

I got the job.

As part of the education and orientation program, the newly hired summer staff met with John Piatt, a PhD student at Memorial University of Newfoundland, who was going to introduce us to the stunning seabirds at Cape St. Mary's Seabird Reserve. A two-hour drive from St. John's and a one-kilometre walk along sheer cliffs brings you to within spitting distance of Bird Rock, the second-largest northern gannet colony in North America. Think raucous, stinking, elegant, goofy, aggressive. Think five-foot wingspan and a blue, sabre-sharp bill. Think tucking wings and dropping like a torpedo out of the sky, disappearing under the ocean surface, dive-bombing fish. Times fifteen thousand. That's Bird Rock. Now surround that by tens of thousands of other seabirds—common murres, razorbills and black-legged kittiwakes. Emerald-green grasses giving way to sheer cliffs. Add whale blows to the horizon and you have Cape St. Mary's. Up to that point in my life, wildlife encounters were mostly an idea; the nature park brought the animals to me. I could smell the sharp, musky scent of the Arctic fox. A great horned owl was taking advantage of the buffet to be found in the snowshoe hare enclosure, the hares' numbers dwindling each day. But Cape St. Mary's was the most spectacular place I had ever seen, an unbridled abundance of wildlife just two and a half hours from my front door.

Seabirds breed in dense colonies on protected islands near rich fishing grounds where they have ready access to food for their young. Cape St. Mary's fit the bill. However, I would learn that oil pollution was already threatening their chances of survival.

In the early eighties, ships travelling off the coast of Newfoundland took advantage of the remoteness of the area and the lack of adequate offshore surveillance coverage to illegally dump their bilge water. The oily waste was killing seabirds on the open ocean, but it was difficult to estimate how many. That was one of the questions John was attempting to answer. Our first encounter with him was at Cappahayden beach en route to "the Cape," where he was recovering dead oiled birds. It was a grim introduction to seabird ecology, but an important one.

When we arrived, John had already begun working and was at the far end of the beach with his dog, Jonah. From a distance, they looked like a pretty standard man-and-his-dog. But as they came closer, both grew to well beyond standard proportions, John at six foot eight and Jonah a solid 55-kilogram southern shore water dog. John was soft-spoken but passionate, his eyes equal parts intensity and humour. Behind those eyes was a mind full of all the things I wanted to know. He was an expert in his field and utterly captivating. I would have followed him off a cliff.

John impressed upon us the global significance of our seabird populations in Newfoundland and explained that of all the birds that get oiled and die at sea, only about one in ten washes up on land. There was a specific set of beaches that he surveyed every month, Cappahayden being one, to determine how many dead oiled birds were washing ashore.

We spent the rest of the day fanning out along the width of these beaches, picking through the seaweed and driftwood and the wrack of plastic detritus—the bits of net and buckets and old buoys that litter most Newfoundland shorelines. We focused our attention on the high-tide mark, where corpses were most likely to remain, collecting and counting dead birds, cataloguing the degree of oiling and other relevant data. If you weren't looking closely, you could easily miss the evidence—a wing here, an eviscerated corpse there, its belly torn open, eaten by a gull or raven. Sometimes the oil was as black as the bird's feathers themselves. Sometimes the bird's identity was disguised by blackened feathers that would normally be white.

Identifying oiled seabirds was an art. I had questions. If you find just a wing, how do you know the bird was oiled?

Smell.

Ah. That deep, pungent, unmistakable odour.

Still others looked almost perfect, just a pale smear on the chest. A little oil maybe, but surely not enough to kill?

John described the delicate balance. The North Atlantic can be survived only if you are built to withstand the cold. Our seabirds are adapted beautifully—the combination of fat and dense down feathers, covered by a meticulously oiled and preened outer layer of feathers. The interlocking surface of feathers maintains an essential barrier to the frigid water. A brilliant design, really—the kind nature produces and engineers attempt to duplicate. But it takes only a spot of oil the size of a quarter to destroy the system. Once the surface feathers are compromised, the insulating barrier is broken and near-freezing water makes it to the skin's surface. The body works overtime to try to create heat, to replace what is being lost. Birds oiled in this way are usually emaciated, having burned all their fat reserves in a last-ditch effort to survive. If near enough to the coast, they will try to haul themselves out of the water—the element they were designed for, now their greatest enemy.

I had no inkling that the oil that was killing seabirds was directly linked to the plastic waste littering the beaches. That plastic *is* oil, with a few chemicals added and subtracted. And that plastic was threatening seabird survival from the inside out, just as surely as oil was doing so from the outside in.

————

My early impressions of wilderness had been shaped by documentary films and glossy magazine spreads, the photographers and filmmakers invisible behind the camera lens. A landscape devoid of humans. I enjoyed the company of others, but wilderness was a separate, sacred thing, a place where there was no evidence that humans existed, or had ever set foot. Remote and unblemished.

I took that sense with me on my first field experience to distant wilderness—in this case, the Torngat Mountains—as a field assistant to a botany grad student named Terry Hedderson.

Located two hundred kilometres north of Nain—the most northerly community in Labrador—it was isolated and difficult to get to. In early July 1986, we boarded the *Sir John Franklin*, a Canadian Coast Guard icebreaker leaving St. John's harbour bound for its Arctic mission, alongside a geography field crew that would also be working in the Torngat Mountains. The ship carried the five of us and all the gear and food we would need for the next six weeks. The trip took several days, the ship staying well offshore until we approached our destination, Nachvak Fiord, an icy finger of the Labrador Current that pierces through the heart of the mountain range for forty-five kilometres. One early morning, I came out onto the deck of the ship, casually lifted my eyes to the coast and gasped from shock—equal parts awe and dread. The word *Torngat* is derived from the Inuktitut word *Torngait*, meaning "place of spirits." I stared at a black wall of mountain, rising well over a thousand metres from the sea to hair-splitting, razor-sharp peaks. There was no vegetation to soften the hard lines, or to suggest survival here was even remotely possible.

Place of spirits. That felt about right. It was sunny and completely still, the mountains' mirror image reflected in the water, suggesting the depth matched the height of the summits. And beyond, rugged peaks as far as the eye could see. The landscape was thrilling and vaguely threatening. The scale of it.

A small Inuit crew from Nain were fishing for char in the fjord. Before departing, they made radio communication with the ship's captain to report that a polar bear mother and her two cubs had been on the beach where we would be landing, but the sound of the helicopter ferrying us and our gear into the fjord had scared them off. We kept an eye out for them as we made camp, carrying equipment, setting up tents, swatting blackflies.

Over the ensuing days the camp fell into a rhythm. The geologists would leave in the morning, sometimes on foot, sometimes by Zodiac. Terry and I would hike to areas of his choosing near the camp—up the valley and on mountainsides—taking moss samples from various habitats and exposures. We both carried small handheld magnifying lenses, an essential tool of the trade, and paper for storing the collected moss. Dressed in layers, because weather conditions would change quickly. While he contemplated moss samples, I followed behind him with binoculars and my copy of *Peterson's Field Guide to the Birds East of the Rockies*. I would check out the low-elevation alder beds, valley meadows and open tundra for species that were new to me—Lapland longspur, northern wheatear. Watch the golden eagle that was seen regularly near our camp. Take in the views that were spectacular no matter where you stood. Reflect for a moment on the immense privilege to be in this place, so wild and difficult to get to. Occasionally I would take out my hand lens and examine a moss, just to be polite.

After three weeks it was time to move farther afield, deep into a valley that led away from the coast and into the heart of the mountains, *Torngat's Lair*. It was too far a commute to base camp, so we packed a tent and supplies for several days.

We had not travelled beyond the view of the fjord before, and the farther we got from it, the more uneasy I became. The ocean was my constant source of reassurance in the otherwise unfamiliar landscape, and its absence from view was disorienting. As we gained altitude, it became cooler, the wispy clouds now a light fog around us. The river valley meadow gave way to exposed rock. We set off at an aggressive pace. There was no commiserating, no pausing to rest or take in the view. Exertion displaced apprehension.

After two days we made it to Torngat's Lair—the dark valley in the heart of the mountains, in deep shadow cast from the surrounding peaks. Terry was briefly elated to have arrived. According to Inuit legend, this was the home of Torngarsoak, the most powerful of the spirits, controlling caribou movements. The place was like a cathedral to an exacting force. Unseen but oppressively felt. It demanded reverence, respect and silence. We obliged. Conversations were brief and held in low voices. I can't remember what mosses, if any, we collected. Maybe the trip merely confirmed that not even moss could survive there. We camped only as long as necessity required, and then retreated.

The hike back felt like a celebration. Our packs and our moods much lighter. Giddy with relief. The interior mountains now behind us, we could glance back and appreciate their stark beauty. The farther away, the less intimidating they seemed—out of Torngarsoak's grasp. And the fjord was in view—we were within range of the familiar once again. We camped in a meadow along the stream that led back to the fjord.

I was woken during the night, startled by what sounded like pigs grunting around the tent. My first thought: *What does a polar bear sound like?* "Terry! Terry!" My urgent whisper didn't wake him up immediately. "What is *that*?" Having been asleep, he had no ready answer. He sat up, opened the tent zipper for a look.

A caribou herd had been making its way down the valley behind us, grazing and advancing slowly, imperceptibly, like the glacier that had carved the path eighteen thousand years before. The grunting, throaty sound of caribou chewing on the other side of the thin nylon tent, unaware. The noise of the tent zipper sent a flare of panic through a thousand caribou, now thundering past us, the herd itself splitting like a zipper around the tent. Miraculously, we were not trampled.

Our return to camp felt like a homecoming—a return to the company of the other three, sharing laughs, trading stories. We told them about the stampede of caribou. They told us about the bear tracks they'd seen a little farther down the fjord, during a day trip in the Zodiac. Just above the high-tide mark.

Many of the days after our return were spent in solitary exploration, when I would set out with my binoculars, camera and field guide and explore the stone terraces of the beach that gave way to stunted alder beds and alpine meadow. Taking in the silence. The scale. Most of all, the stillness. Nothing moved, except the clouds and the caribou.

I was also discovering that, for all the expansiveness, the Arctic landscape held great subtlety; upon close inspection, there was a whole other world being inhabited on an entirely different scale. Walking toward the backshore on one of these adventures, I almost stepped on a tiny shorebird, drooping its wing as though injured. I recognized this diversionary tactic common in shorebirds, designed to distract and deflect potential harm away from a nearby nest, and backed off immediately, not wanting to cause undue stress. From a comfortable distance I watched for a long time. No longer considering me a threat, the bird went about its business, scurrying this way and that, poking amongst the rocks and meagre vegetation, turning on a dime, flitting and landing. I was treated to every angle and aspect and could take in the identifying field marks at my leisure—the length and shape of the bill, the leg colour, wing length, the extent of the buffiness on the chest. Eventually I identified it as a Baird's sandpiper. The place was so remote that the edition of my field guide didn't include Labrador in its range map. Not too surprising—it felt like a place beyond human range as well.

However, not far from the nest I came across an ancient pile of stones on a terrace behind our campsite. Covered in black lichen, it had likely been undisturbed for centuries. I peeked between the stones and saw a human skeleton, accompanied by pieces of wood and other items, carefully placed. It was a grave. I didn't know who the inhabitant might be, or the significance of the objects that lay beside them, but I understood that it was sacred and private and I had no business there. I bowed my head in a small gesture of reverence and backed away. I didn't mention it to the others, and I didn't return myself. It was easy to imagine that we were the first people to set foot in this apparently untouched wilderness. But of course Inuit had been living and dying here for thousands of years, leaving barely a trace. Their tools, their boats, their clothing, their food were of the place. And to it they returned.

I had come to Nachvak Fiord with the naive impression that we would simply be superimposed over the landscape, having no effect on it or each other, our impacts somehow edited out. But the fact is, you cannot consider humans and wilderness as separate entities. Our garbage burn pile—the Hard-Dee freeze-dried food packets and, despite advice to the contrary, disposed feminine hygiene products. Meat and cheese wrap. Food waste. The smell of burning plastic and organic material wafting into the pristine air. The wilderness of my imagination was in some ways the landscape inhabited by ancient cultures—living close to the land, their presence "erased" by the natural forces of microbes, wind, sun and time. The unmarked grave was a poignant reminder of that. But in the modern era, much of the human race has embraced conveniences that have left a trail of pollution on a geological scale. So much so that the term *Anthropocene* has been coined to describe a new epoch—starting in the 1950s (when plastic was being produced on a commercial scale) and marking the beginning of significant human effects on the earth's climate and the environment. Humans are an integral part of the landscape and always have been. It is the impact we have that counts.

Several days after I encountered the plastic diaper on Killiniq Island, the *Sea Adventurer* sailed into Nachvak Fiord. It was the first time I had laid eyes on the place since 1986, exactly thirty years earlier. During that first six-week trip, we didn't see a single polar bear. This time we spotted no fewer than fifteen dotting the shoreline on rock outcrops and terraced beaches along the length of the fjord. There were likely more, hidden behind features that could not be seen from the ship. Some were travelling slowly, but most were lying still, conserving energy, waiting for the ice to form—a platform from which to hunt for seals. There were plenty of harp seals around. The seal population had exploded after the ban on seal products was imposed by the European Economic Community in 1983—and in response, so had the regional polar bear population. But the seals would remain out of reach until freeze-up. The polar bears had not had a substantial meal in months.

In the summer of 2013, an environmental organization called the Sierra Club had launched a seven-person hiking expedition into the Torngat Mountains that started in Nachvak Fiord, at the exact location of our camp back in 1986. The group had turned down the opportunity to be escorted by an Inuit polar bear guard, relying instead on the protection of an electric fence around the perimeter of the camp—a decision they would soon deeply regret.

Matt Dyer, an avid outdoorsman from New York, was one of the hikers. On the second morning at camp, Dyer woke to the sight of a bear silhouette through the thin wall of his nylon tent. He had just enough time to yell a bear warning before he was grabbed by the head, ripped free from the nylon tent and dragged toward the beach. The trip leader responded with two flares. One that scared the bear enough to drop his prey; the second to deter the returning bear, which was not ready to give up its meal. In the retelling of his harrowing tale, Dyer recalled the awful sounds of bones crushing—his jaw, his neck. Or was it his skull? The slick wet of saliva running down his face.

The polar bears we encountered during my second visit to Nachvak Fiord were clearly hungry, and at that density it was deemed unsafe to make a landing anywhere in the fjord. It was a different world from the one I had first visited.

The disappearance of ice, the emergence of plastic garbage—all of this is arising from our dependence on petroleum and other fossil fuels: products that are unequivocally driving climate change, and are solely responsible for the plastic crisis.

———

Plastic has always needled at the edge of my consciousness.

Memorial University professor and plastics researcher Max Liboiron describes coastal plastics as having the accent of the place in which they are found. This rings true. Hunting and fishing are central to the culture in Newfoundland, and people's relationship to the environment has largely been a practical one. If you didn't need something, you left it behind. The attending detritus—shotgun shells, abandoned and lost fishing gear, salt beef buckets, anchor buoys—was as much a part of the beach landscape as the rounded stones and storm-strewn seaweed. For most of my life it was simply an irritation., a. An insult to the landscape that I loved. The diaper I encountered at the tip of Labrador in 2016 deepened that unease, but I was still mostly saddened by the sight of it. By the circumstances that led the bear to eat it. I didn't know how toxic the meal was, or the true nature of the crisis it represented.

When it came to marine threats from man-made materials, nets were my only real concern. Fishing gear was always a worry—nets lost or abandoned and washing ashore with the occasional seabird enmeshed. Farther down the coast from Killiniq Island, a nylon gill-net lay half-buried in the sand. A thick-billed murre was twisted and rotting, entangled and drowned in the mesh. Had it washed off a boat deck in a heavy sea? Been left carelessly close to the shoreline in a heap, only to be carried out by the next high tide? Tossed? Curtains of lethal mesh drift aimlessly in the water column, taking any fish, seabirds and marine mammals in their path. The animals trapped and dying before being released in decaying pieces from the mesh. Making room for more. These ghost nets were a serious problem, but at least it was a problem with a shape, with boundaries.

The diaper on Killiniq Island was a sign that those boundaries were being erased.

Chapter 3

The Heart of the Labrador Current

Most people who live near the coast know the ocean from the perspective of what is in front of them—the body of water that exerts specific forces in response to local conditions. For me, that is the Labrador Current.

It has informed my impression of what the ocean is capable of and can provide. The waters made frigid by glacier and sea ice melt, and carrying with them an ecosystem that once seemed boundless in its wealth of marine life—seabirds, whales, seals and fish—more than enough to support both the creatures that feed on them and the livelihoods of the fishing communities that dot the coast. But it's become clear to me how fragile the balance is. The Labrador Current is a microcosm of the wider crisis that has developed in the oceans around the world.

It has flowed by my doorstep my whole life; it is the reason Europeans settled in cold, foggy, rainy Newfoundland in the first place. It is why you can grow only the hardiest of root vegetables here, and salt is considered a food group. It's why the phrase "It's a nice place to visit, but you wouldn't want to live there" exists. It is why there is such a rich history of music and storytelling. It is why subtitles are provided on national newscasts when Newfoundlanders are interviewed; why the "Newfie joke" exists. It is why, in the midst of the horror surrounding 9/11, when planes were grounded and American citizens were stranded here, lifelong friendships were forged and scholarships established to honour the actions and response of Newfoundlanders—why the hit Broadway musical *Come from Away* was created. It is why we pull together and help each other. It is why I, like so many Newfoundlanders, tried to live somewhere else but it never quite took. In a nutshell, the Labrador Current has ruined our climate and shaped our culture. And it is in no small part responsible for who I am.

Let me explain.

The Labrador Current is often portrayed as a ribbon of fast-flowing cold water that snugs against the coast of Newfoundland and Labrador, extending to the continental shelf, wrapping around the Grand Banks of Newfoundland—an extension of the continental shelf—before petering out, outmatched by the warm waters of the Gulf Stream. Like all things oceanographic, it is more complicated than that. The ribbon is, at the very least, frayed. One strand flows along the coast to the edge of the continental shelf, crawling at a pace of 20 kilometres per day. Another, much faster strand follows the edge of the continental shelf at a brisk walking pace of about 80 to 100 kilometres per day. The third strand extends into very deep water (three to four kilometres below) and moves only a few kilometres a day. A snail's pace? The third strand forms part of the Atlantic sub-polar gyre, an area of ocean circulation centred just south of Greenland. It is fed by a mixing of warm and cold currents in the North Atlantic and has important implications for global climate. All three strands of the ribbon are characterized by high freshwater inputs from sea ice, Arctic river runoff and icebergs. While the third strand is busy influencing global weather patterns, the faster-flowing strands provide the conditions for a rich marine environment around Newfoundland and the Grand Banks. These two strands will be lumped together here for simplicity's sake.

A large portion of the cold waters of the Labrador Current originate in western Greenland, where the glacier ice cap that covered all of North America eighteen thousand years ago still remains. Glaciers are formed from the compaction of snowfall over thousands of years. For this reason, glaciers and the icebergs they calve are also made of fresh water. Most of the icebergs in the Labrador Current originate in Ilulissat Icefjord, located 250 kilometres north of the Arctic Circle. Here, the Jakobshavn Glacier flows imperceptibly, like an ancient, arthritic river. Pillars of ice gouging, breaking, cracking and folding over the rough terrain in its path. Once it reaches the coast, the massive ice sheet spills into the ocean, breaking off, or "calving" mountains of ice into Disko Bay.

From Disko Bay, the path of the current takes the shape of an inverted U. It hugs the coast of west Greenland then merges with glacial waters from the Canadian Arctic and turns south along the continental shelf, forming the Baffin Current. Icebergs can spend years in Baffin Bay before finally being released into the Labrador Current, where they are pulled along by its force. As the icebergs travel, they weather, break and roll, forming massive ice sculptures with colours of white, blue, purple. Patterns emerge from running rivulets of water on the melting surface. There is a thunderous roar when pieces break off, destabilizing the tenuous balance and causing the whole thing to roll until it finds a new equilibrium, creating tsunami waves in its wake. The rule of thumb for floating icebergs is that only one-tenth appears above the surface of the water; the other 90 percent is hidden beneath the surface, the mass and power imperceptible from above. When these things are on the move, you'd better not be in the way.

The relationship between the Labrador Current and iceberg movement has earned it the nickname "Iceberg Alley." The formation of winter sea ice in Arctic waters interrupts the icebergs' migration, temporarily holding them in place. But once icebergs are released from sea ice in spring, they are free to move in open water with the Current. They are as treacherous as they are beautiful. Over five hundred ship collisions have been recorded around Newfoundland and Labrador, a testament to the ruthless power of the ice. The vast majority of these have occurred on the Grand Banks, where Newfoundland's offshore oil platforms are clustered.

By August, most icebergs off the coast have melted. The entire journey from Disko Bay to the last wisps of the Labrador Current takes years, but once they are released from sea ice, their demise is certain.

The high content of sea ice and glacial meltwater in the Labrador Current gives it some unique qualities. The surface water temperature is at its coldest in spring (0°C on average) and has a lower salt concentration than most oceanic waters. Cold water dissolves more oxygen than warm water, so it is also very oxygen-rich. Upwellings from the deep ocean that collide with the continental shelf bring nutrients to the surface that mix with the Labrador Current. When you mix highly oxygenated water with nutrients and light, you have conditions for massive marine productivity, starting with the base of the food chain—phytoplankton. Phytoplankton are tiny, free-floating marine algae, from the Greek words *phyto*, meaning plant, and *plankton*, meaning to wander or drift. When the conditions are right, plankton can have explosive growth, called blooms, which sometimes spread over several kilometres and last weeks, making the water murky and dense and difficult to see through.

April can bring the right conditions, which I witnessed when I did my first open-water scuba dive during a phytoplankton bloom one year. It is the month with the coldest ocean water temperatures, when sea ice has melted but icebergs are still around. The same month that an iceberg took the 'unsinkable' Titanic down in 1912. I wore a wetsuit that proved to be too large for me, with gloves, boots and a hood. My unprotected face froze; I couldn't feel my lips but managed to hold my regulator in place. My instructor was close by my side for reassurance and guidance. We were not going any deeper than thirty metres, but it wasn't long before the water pressure caused my wetsuit to contract around my body; my boots compressed, conforming to the actual size of my feet. One fin fell off and I reached back to grab it before it floated away. The second fin followed suit, and I turned to grab that too. I found myself standing on the ocean floor, a fin in each hand—and, to my great consternation, no instructor in sight. I stood there, staring, straining to see past my hand. I could just barely make out my feet and a few of the closer rounded boulders fading away before disappearing into a thick, green-hued fog. Above, nothing. We hadn't covered this scenario in class. Do I stay where I am? Do I go to the surface? I decided to follow best practices for being lost in general and stayed put. It wasn't long before my instructor appeared out of the murk, staring at me quizzically, brow furrowed. Even through his mask, the question on his face was clear: Why are you standing there wearing your fins on your hands? The point is, he hadn't gone more than a few metres before he completely disappeared. That's how dense the bloom was.

Phytoplankton is the first spark of life in any marine food web. It is grazed on by tiny marine animals called zooplankton. This is where the web gets tangly and complicated. In a perfectly linear situation, small plankton feed on larger plankton, which are then eaten by small fish, feeding bigger fish, and up the line to seabirds and marine mammals. These steps are referred to as trophic levels and many animals feed at multiple trophic levels or skip trophic levels altogether. For example, the blue whale—the largest animal ever to have lived on the face of the earth—feeds almost exclusively on krill, a type of zooplankton, at the rate of around sixteen tons a day during the spring and summer. But when a blue whale dies, the tables turn and zooplankton feast on the carcass. Who feeds on whom can vary depending on the age and stage of the creatures involved.

In the northwest Atlantic, the animal at the centre of this complex food web is the capelin, which itself feeds on the abundance of copepods and other zooplankton. It is a small, slender cold-water schooling fish about the length of a pencil. It is their small size and abundance that make them the perfect prey—easy pickings for just about everything else. Capelin spend the winters offshore, migrating en masse in spring either to sandy and pebbled shoals on the Grand Banks or to shallow waters near the coast, preparing to spawn on beaches when the temperature is just right, around 6°C. The males arrive first, waiting in place, ready for when the females show up. The females hang just off the spawning beaches, waiting for the right conditions. When the water temperature hits its mark, the females will swim toward the beach, getting carried up by a wave, with one or two males in pursuit, vying for position. Together, they fan their tails, making a depression for the eggs, which the male then fertilizes with milt. The female lays thousands of eggs in one spectacular attempt to reproduce, before leaving the coast to die. Males will make several runs at the beach before, exhausted and beaten from the rocks, they die too.

While capelin are single-minded in their purpose, there are other agendas afoot. The beach crowded with families—children squealing in squeamish delight, capelin rolling and sliding against their bare legs, the bottom obliterated by their packed density; people scooping up fish with bare hands or dip nets, dumping them, writhing, into buckets. Dead, spent capelin collected to fortify the soil in vegetable gardens. Gulls scream and dive, plucking live and dead fish out of the water, swallowing them whole. By the time it is all over, the beach is covered by a thick, spongy layer of eggs and countless thousands of dead capelin.

It is no exaggeration to say that coastal life is synchronized with the arrival of capelin for survival.

Seabirds. Whales. Cod. People.

So much abundance can give the mistaken impression of having no limits, being beyond our capacity to influence. But the fate of the great auk serves as a cautionary tale.

The great auk—the supersized relative of the little auk—was once abundant in the North Atlantic. It stood about eighty centimetres tall (just under three feet), and with stubby wings adapted for swimming, it was unable to fly. It was black and white, and although completely unrelated, it resembled a penguin in every way. In fact, its Latin name is *Pinguinus impennis*, and the birds were referred to as penguins in the early days. But when explorers later discovered the flightless look-alikes in the Southern Ocean, the name was applied and eventually completely transferred to the group we now know as penguins.

European fishermen and explorers were well familiar with the great auk. It nested in dense colonies on select islands in the North Atlantic, from Newfoundland to Norway. The largest known colony was located on Funk Island, sixty kilometres off Newfoundland's northeast coast. Eight hundred metres long and thirty metres wide, it was little more than flat, bald granite protruding out of the ocean, part of an area of shallow shoals that had rich local productivity. Coupled with its distance from shore and any terrestrial predators, Funk Island was the perfect location for an auk colony. The size of the population that flourished there is not known, but for the sake of scale, they would likely have numbered in the hundreds of thousands. Outside of the summer breeding season they dispersed, largely to the Grand Banks. Great auks were so plentiful they were used as a navigational aid: sailors bound for Newfoundland or the Grand Banks fishing grounds knew they were getting close when they started to encounter great auks on the water.

On the colony, these densely packed, flightless birds were a welcome source of fresh meat and eggs after transatlantic voyages by European fishing fleets visiting the Grand Banks as early as the 1500s. Their feathers later became a highly valued commercial commodity, for use in mattresses and bedding. By the mid-1700s crews were stationed on the island for the summer, with the sole purpose of harvesting auks. They were herded into stone corrals, where they were killed; some were boiled in water to loosen the feathers from the skin, and others were burned, their fatty carcasses fuelling the fires. This kind of wanton exploitation was happening all over the species range. Even George Cartwright, a British army officer, explorer trader—and no stranger to wildlife pillaging himself—warned in 1785 of the demise of the great auk. "A boat came in from Funk Island laden with birds, chiefly penguins [great auks]," wrote Cartwright. "But it has been customary of late years, for several crews of men to live all summer on that island, for the sole purpose of killing birds for the sake of their feathers, the destruction which they have made is incredible. If a stop is not soon put to that practice, the whole breed will be diminished to almost nothing."

Of course, George Cartwright was right—or almost right. In 1844 the last pair of great auks was killed on Eldey Island off the coast of Iceland by a couple of fishermen, the last egg crushed under a boot. And thus "the whole breed" was diminished to absolutely nothing. The great auk was simply too easy and convenient to kill for its own good. Its extinction serves as a stark reminder of how fragile the natural world can be—when a seemingly boundless resource is subjected to our all too human desire for what's easy.

Fortunately, the great auk has living relatives that still thrive in the waters of the Labrador Current today. The Latin name is Alcidae, but informally the same family of birds is known as "the auks"—puffins, murres, razorbills, guillemots, dovekies. These birds bear a strong resemblance to the great auk in appearance and share the same appetite for capelin. And like the great auk, they gather by the tens and hundreds of thousands on offshore islands in the summer to breed, synchronizing the hatching and rearing of their chicks with the timing of the massive inshore run of spawning capelin.

The living auks are not flightless, but not far from it. The short wings that afford them the ability to dive deeply and move with swift precision underwater comes at the expense of agile and efficient flight. They prefer to do as little of it as possible, and nesting colonies are selected for proximity to rich waters with access to inshore capelin. There are ample locations where these conditions are met, particularly along the east coast of Newfoundland—Funk Island, Baccalieu Island, the Witless Bay Islands, Cape St. Mary's, to name a few.

The auks are joined by a host of other breeding seabirds on these islands—black-legged kittiwakes, northern gannets, Leach's storm-petrels. These colonies support millions of seabirds and host the largest Leach's storm-petrel colony in the world, the largest and second-largest Atlantic puffin colonies in North America, the largest common murre colony in North America, the second-largest northern gannet colony in North America—the list goes on. In addition to the local breeders, there are the shearwaters that are common offshore in summer but venture close to the coast to feast on capelin. Great shearwaters, sooty shearwaters—both a common sight for fishermen, with local names like *bawk* and *hagdown*, for reasons long forgotten. I had always thought of them as "our" birds. But they are borrowed from colonies thousands of kilometres away—on remote islands in the Southern Ocean, off the south coasts of Africa and South America. They make the epic journey, under mysterious guidance, to feed in these rich waters during the austral winter.

As for Funk Island—the absence of the great auk left spaces for its smaller relatives. Common murres breed shoulder to shoulder in numbers estimated at close to 400,000 pairs. Scattered among them, thick-billed murres and razorbills. In the days of the great auk, there was no soil for puffins to burrow into; they couldn't have survived there. In what seems like a parting gesture, the discarded carcasses of the great auks left just enough organic material behind for soil to begin to form, and eventually a safe haven for puffins to breed.

The migration of the humpback whale is not as long as that of the shearwaters, but the circumstances are equally impressive. These mammoth whales grow to about fifty feet, weigh fifty tons and spend most of the year in warm waters where they breed and give birth, all the while fasting. They travel north for one reason and one reason only: to feed. They have to gain enough weight to replenish and sustain them for the subsequent eight months without food. Humpbacks migrate en masse, and are joined by other capelin enthusiasts—fin whales, minke whales, white-sided and white-beaked dolphins. And of course, the cod.

Cod feed on all kinds of things, depending on their age and the time of year. I spent a winter, desperate for employment, analyzing cod stomachs for the Department of Fisheries and Oceans (DFO). Barrels of formalin-preserved stomachs waited to be sliced open, followed by the tedious task of identifying, counting and measuring everything they contained. Guessing which bits were once attached to other bits. Counting bumps on a shrimp's rostrum, to identify the species. Then counting the shrimp themselves. It was interesting for the first week.

Sometimes the stomachs were almost empty—the good ones, from my perspective if not the cod's. Other times they were an extravagant mishmash of shellfish, krill, marine worms, tiny octopus, sea cucumbers, sea squirts, other cod, even a seahorse once—anything that wasn't nailed to the ocean floor. The contents can occasionally be more extravagant and unexpected. Dr. Wilfred Grenfell, a medical missionary who served the coasts of Newfoundland and Labrador from the 1890s to the 1920s, once claimed he was given a leatherbound encyclopedia that had been found in the belly of a cod. Flash-forward a hundred years and it gets stranger still. In 2020 a fisherman was gutting a five-kilogram cod near his home in Norway. Amongst the slurry of partially digested food, he found—to his great astonishment—a vibrating dildo. We didn't find anything quite as stimulating, mentally or physically. There was the occasional bottle cap or cigarette butt. One of my fellow lab technicians did find a fully intact Coke can once. It was filled with perfectly preserved sand lance skeletons, protected by the can from the digestive action of the gut, which would otherwise have pulverized and digested them. The can itself, indifferent to the stomach's effort to digest.

Everyone's favourite barrels were those with cod that were caught in the summer. Large, rigid stomachs full of capelin—easy to count and easy to identify. And solid evidence that cod, like everything else, chased capelin on their way inshore to spawn.

The amount of capelin required to support this whole system is staggering. Each year about three million tons is consumed by all and sundry. In the eighties, when my eyes were first cast seaward, the estimated capelin biomass was in the millions of tons—all of it possible because of the Labrador Current. There was plenty to go around, for both the fish and the fishers who depended on it. Growing up in a place of such abundance, it was easy to be seduced into believing the supply was endless. That nature was something that existed outside of us and our influence.

The abundance created and sustained by the Labrador Current is the sole reason Europeans showed any interest in Newfoundland. The history of the cod fishery in Newfoundland goes back to 1497, when John Cabot set sail with an English crew in search of a western trade route to Asia. Cabot encountered Newfoundland, believing it to be the east coast of Asia. And while missing his mark by some considerable distance, he had "discovered" one of the richest fishing grounds in the world.

Word spread throughout Europe, and by the early 1500s there were fleets from Portugal, France and Spain involved in the inshore fishery. By the mid-1500s the Grand Banks stocks were discovered and the offshore fishery ensued. During the summer months, fishermen would build rough premises, wharves and stages in order to process the fish—either pickling them in barrels with salt, or salting and drying them on wooden frames called flakes, then packing them to ship back to Europe at the end of the fishing season.

The first attempts at permanent settlement were a disaster. The British tried their hand in the early 1600s in locations around the Avalon Peninsula, which lies south of both England and Ireland, along the same line of latitude as Brittany, France. And while Brittany has a wine region, the town of Cupids on the Avalon Peninsula had barely enough soil to grow root vegetables. They couldn't grow enough hay to sustain their animals. The Labrador Current made for miserable weather and a short growing season—conditions that rendered the land too miserly to sustain the inexperienced—and many didn't survive. Fishing was a hard, dangerous job at the best of times. The Newfoundland fishery attracted those who had nothing to lose.

The Irish started joining the British fleets in the late 1600s, participating in the Newfoundland fishery as indentured servants. After three years of service they would earn their independence and could start fishing for themselves, settling in remote coves and harbours around the island that provided access to local fishing grounds. Nothing encourages co-operation and a sense of community like hardship, and there was plenty of that to go around. If the fishing season didn't go well, there was the risk of starvation. With a sudden change in wind direction, storms could be whipped up on a dime. And desperation sometimes led to risk-taking heading out onto sea voyages they may not return from. Often, they didn't. The family left behind in grief, but also peril; both loved ones and livelihood taken at once.

Isolation meant little or no access to medical attention . Or exposure to other ways of life. Time stood still. If one family was lacking, neighbours helped out. Together, they built boats, wharves and flakes, mended nets, dug gardens. People were resourceful and pulled together. A good sense of humour helped too. And then there was the kitchen party—people gathered in kitchens to play instruments, sing songs, tell stories. The remoteness and isolation bred unique words, sayings, dialects and accents. With the passage of time these deepened, until you could tell what region of the province someone was from by their accent, how they described the weather, what they called a sea duck. This is still true today.

Growing up we had a housekeeper, Mrs. Wade. She helped my mother out with light housework and unofficially kept an eye on us. She was slender, with the small pot-belly of a woman who had given birth to six children, and she always ate her lunch standing up, leaning against the kitchen counter. Always, a tinned tuna fish sandwich on white bread and a cup of tea. Her left elbow bent, wrist hanging limp. Mrs. Wade lived in the fishing community of Flatrock, just twenty minutes north of St. John's. Her husband was a fisherman; she had known many people who had been lost to the sea over the years, and she knew that cod will eat just about anything. Mrs. Wade didn't eat cod. She told us that when people were starving and desperate, they would resort to eating lobster—the scavengers of the sea. It was a humiliation, and the shells were carefully hidden in the garbage so no one would know.

My mother often drove Mrs. Wade from our house to her home in Flatrock—a half an hour and a world away—with some combination of us kids in the back seat, tagging along either out of necessity or curiosity, depending on the age. There was something exotic about the place to us suburban kids—the houses, mostly saltboxes from the previous century. Not in neat rows, but peppered between areas of exposed granite, reached by crooked lanes that yielded to the shape of the landscape. The wind blowing in the harbour, right there. The boats, the slipway, the piles of nets—the heart of the place. On one of those drives I remember my mother remarking how strange it seemed, that the windows of the houses were small and didn't face the view of the ocean, the larger windows favouring the view of the road instead. Mrs. Wade explained what to her was obvious: the worst storms came off the sea, and it was best to protect against the ruthless cold, wind and battering rain. Besides, you lived your life on the edge of the ocean, if not on it. Who needed a view from the kitchen window?

To come from St. John's is to be considered a faux Newfoundlander by those who live anywhere else on the island—a Newfoundlander-lite. The city supports a more diverse economy, with access to education and medical care and fast-food outlets—a rhythm of life not dictated by the vagaries of the sea. Much has changed in the past fifty years or so, but the attitude has not. And while it's true that most "townies" don't depend on the ocean in this elemental way, we have retained the casual reliance on each other. It was something I wasn't really aware of until I moved to British Columbia in the early nineties. I hadn't lived anywhere else in the country, although I had done lots of travelling in other parts of the world and was accustomed to moving around in other cultures. But I was not expecting to have to make significant adaptations with the move; it was Canada, after all, and I was looking forward to the change.

My first months in Vancouver were lonely. I was working on my master's at home and knew nobody, only encountering other people at the supermarket. One day I stood over a display of Jerusalem artichokes, next to another woman who was holding one in her hand, feeling the firmness, turning it to assess the colour and texture. I was curious, so I asked, "What do you do with that?" She looked at me blankly, so I tried again. "How do you cook it?" (*Maybe she doesn't speak English?*) She put the vegetable in her cart and turned away. And left me standing there, feeling awkward and none the wiser about Jerusalem artichokes. Another time, I walked into a Mexican furniture store downtown. The owner was behind the counter and as I walked around the store, I called over my shoulder, commenting on the weather, the beautiful carvings, the fabric textures, whatever. He responded warmly and unselfconsciously, and said finally: "You aren't from around here, are you?"

"No, I'm not. How did you know?"

"You are more like a Mexican than a Canadian."

I did feel as if I'd moved to a foreign country, from one with a leaner economy, where a sense of community was bred in the bone— essential for survival. Maybe more like Mexico.

No one locked their doors in my grandparents' day. My grandfather had a car, a sedan with a large back seat. He started to notice signs of habitation when he got in it to go to work in the mornings—a worn hat, or unexpected footprints. Someone, he realized, was sleeping there at night. He put a blanket on the back seat.

British Columbia was beautiful, but I didn't belong there. I was forged from the harshness imposed by the Labrador Current and all it conveys. I moved home and started doing office jobs part-time while my growing family was young. Eventually I landed a comfortable federal government job with decent pay and benefits. But financial security was never a big motivator for me; we have one life to live and I didn't want to spend mine behind a desk. After fourteen years I returned to fieldwork. The environment had been undergoing formidable changes while I was changing diapers—the plastic, disposable kind. And as soon as I figured my children were old enough to do without me for days at a time, I returned to the field and my other passion, seabird conservation. In 2008 I was hired by the provincial government to manage the Witless Bay and Baccalieu Island seabird ecological reserves.

Two

The Fragility of Marine Life

Chapter 4

Quiet at Witless Bay

I was shocked by the nocturnal silence that greeted me on my first visit to Gull Island in over twenty years. I'd packed earplugs, remembering the incessant racket of seabirds from my original field season there in 1985. But much had changed since then.

Gull Island is one of the group of four islands that make up the Witless Bay Seabird Reserve about a kilometre off the coast, spreading along the Avalon Peninsula's southern shore. This reserve, just forty-five minutes south of St. John's, is home to the largest Atlantic puffin colony in North America and one of the largest Leach's storm petrel colonies in the world. Add the other usual suspects in a North Atlantic seabird colony—common murre, razorbill, black guillemot, black-legged kittiwakes—and you have close to a million seabirds vying for limited space along cliff ledges and in the soft turf.

The log cabin that served as the research station in the early days had been replaced by a plywood version on the same footprint. It had a steep slope with a peaked roof designed for capturing light energy by solar panels, stored in battery arrays for powering video cameras, computers, cellphones. Tour boats sailed by at regular intervals all day long, broadcasting bird facts and jokes by rote as they passed the cabin ("and this is the breeding season for the biologists too!"). Beneath the cabin, batteries, some plywood and sections of plastic PVC pipe once used for a puffin study had been left behind in a big pile by a former grad student.

Some of the most obvious changes on the island were due to the federal moratorium on cod fishing that was announced in 1992, when it became clear that cod stocks had crashed to a level that constituted "commercial extinction." The complete closure of the Atlantic cod fishery was meant to be a short-term measure, to allow cod populations to recover. This was a crushing blow to the culture and economy of outport Newfoundland that had depended on the cod fishery for five hundred years. It was hard to imagine, but it happened—all fishing gear came out of the water. Thirty years later, the cod stocks are yet to recover and the moratorium remains in place.

There was a silver lining for the auks, though. The puffins, murres and razorbills that had previously laden fishing gear as bycatch were no longer in peril of drowning. In the ensuing twenty years, auk populations had increased. This was most evident for razorbills, the most sleek and elegant of the Atlantic auks and the closest living relative of the extinct great auk. Their black face and thick black bill, shot through with two striking white lines—one from the bill to the eye, intersected by the other wrapping around the bill, like the bridge of a pair of glasses. With the air of an intellectual, razorbills are locally called tinkers ("thinkers" if the h's were pronounced). Nesting in hidden rock crevices and rarely seen on the cliffs or in the water in the 1980s, they were relatively plentiful by 2008. To this day, they can be seen basking and socializing with each other in the comfortable company of murres on rocky outcrops and in the water near the colony.

But it was an entirely different story for Leach's storm-petrels.

———

I knew little about Leach's storm-petrels in 1985, but I was eager to work on any seabird colony. I managed to get a job as a master's student assistant for Libby Creelman, who was going to conduct her research on Gull Island. She had worked on birds in southern California, but decided to trade sun-drenched bird islands for the lure of the Atlantic puffin. And who could blame her? I went to her office to be interviewed, meeting her for the first time. Slender, with long brown hair, soft brown eyes and model-high cheekbones, I would have been a little in awe of her if I hadn't been so nervous. She had a slow, thoughtful way of speaking and looked directly at me. Not in a challenging way, but in a way that said she was paying close attention: Are you someone I can live alone with for three months on a remote seabird island?

I initially thought she was studying another seabird, the black-legged kittiwake, but it quickly became apparent that hers was a puffin study. Puffins have an innate charm and I was not immune to it. This news added a layer of thrill to my already heightened nervous energy—a combination that sometimes causes my brain and body to disconnect and can be lethal to a successful job interview. I could not remain still; I dropped things. I don't know what I said. *"I love puffins!"* I hope not, but probably. I left her office without my folder of important information. I thought I had utterly ruined my chances. Luckily, she appreciated my enthusiasm.

Despite the proximity to St. John's, I had never been here before. This was before commercial tour boats started operating and there was no easy way to access the islands. I had never seen a puffin before, or been so removed from civilization (it was the summer before my trip to the Torngat Mountains). I had never experienced anything like this.

Libby had arranged for a local fisherman from Witless Bay to bring us out. We began to encounter large rafts of birds sitting on the water, diving or skittering over the surface as the boat approached. And as we drew nearer, the racket of the colony: Kittiwakes, complaining their own names in a high-pitched nasal *Kitti-WAKE! Kitti-WAKE!* Murres erupting in a series of guffaws like old men laughing at their own jokes. Thousands of birds greeting, warning, sparring, amplified against the rocky cliffs. The deep, penetrating funk of all that guano—its own force. At the last moment, the whirr of hundreds of puffin wings, as they cleared the hillside in a panicked flight over our heads. I was astonished but tried to remember to keep my mouth closed.

The boat butted up against a natural rock shelf, which served as a landing under good sea conditions. We then started unloading supplies—field equipment, fresh food, tinned goods, stove fuel, and enough water for drinking and washing to last the first three weeks. All the while being careful not to step on the slippery kelp, passing the awkward and heavy cargo in time to the rhythm of the swell, resisting the tug on our boots that threatened to pull us in. The sounds of strained grunting, the wooden thud of the boat against the rocks, the sucking of water around our boots. It was a tense and precarious business, and a relief when everything was above the waterline. We took a few moments to catch our breath before the arduous hike up the steep and unstable hillside, which was honeycombed with puffin burrows. Straining under the weight of five-gallon water buckets, one in each hand. Taking care not to collapse the delicate turf excavated from below. The occasional growl of a disgruntled puffin underfoot. Overhead, gulls took careful aim, dive-bombing, screaming and showering us with guano. The open, grassy hillside gave way to a narrow, wooded path that led, finally, to the log cabin.

Everything was damp. Dense fog clung to the grass, the air, our skin. For the first month it broke only occasionally, always at night. The lights of Witless Bay just visible, offering a reminder that the rest of the world was still out there. But each morning, one look out the window confirmed what you already knew in your heart. Limitless drops of condensation, robbing you of depth perception. The fog was back.

Puffins, like all seabirds, spend their lives on the open ocean, coming to land only to breed for a few short months in summer. Built exquisitely for a life at sea, they shed the colourful plates on their bill in fall and get down to the serious business of surviving a winter on the North Atlantic. Their wings are short, designed for swimming efficiency. They can dive to depths greater than a hundred metres searching for capelin and other small fish, leaving behind a trail of air bubbles like fairy dust. Their feet are set far back on their bodies, rudders for agile manoeuvrability. Under water, they are elegant and masterful. Not so, above. The short wings are almost useless for flight, the wingbeat fast and frantic, just a few beats away from falling out of the air altogether.

In late winter the bright blue, yellow and orange plates are replaced, looking resplendent, proclaiming, *Look at me!* Fresh feathers like a new coat of paint. Summer is all about courtship, breeding and raising young. Barring disaster, a pair of puffins will remain together, sharing the same burrow year after year. Singles search for a mate and come to land to lay claim to an unoccupied burrow if they're lucky, or a piece of turf to dig a new one.

The first few weeks are spent getting reacquainted, clapping bills together, bowing heads to the ground, re-establishing their bond after the full winter apart. At the same time, they are also busy digging and repairing their burrow, using their colourful bills as trowels, small flame-orange webbed feet as shovels, flicking dirt in small sprays out of the entrance. A rudimentary nest of a few strands of grass is built deep within one chamber of the burrow, and in it a single egg is laid. A second chamber is used as a latrine. Both parents take turns incubating the egg and feeding the chick until it is ready to head out to sea for the first time by itself, in August. Libby was interested in how the labour of chick rearing was divided between the parents.

She had already selected thirty burrows that we would be watching, all of them marked with bright-orange tongue depressors and each bearing an identification number. Libby had caught at least one member, if not both, of each puffin pair the season before, using noose carpets—a series of monofilament nooses attached to a sheet of mesh that was secured to the ground with pegs at the entrance to the burrow. Monofilament fishing line was perfect for the job—a plastic designed to ensnare with a fine but strong line, almost invisible. The idea was that when a puffin landed, or stepped out of its burrow, it would get snagged by the foot in one of the nooses. When it attempted to escape, the noose would pull around the foot, trapping it on the spot. Libby would then rush in, hold the bird securely, remove the noose and place a metal band along with a unique plastic colour band combination on the bird's leg, so that individuals could be readily identified.

We were to monitor the comings and goings at each marked burrow from dawn to dusk for the entire season—about sixteen hours a day, divided into four-hour shifts. We recorded activity for each bird. Early in the season it was simply a matter of the puffins changing shifts to incubate the egg, with a little bill clapping as greeting at the burrow entrance. The first fish delivery to a burrow heralded the hatching of a hungry chick. It was important to capture this moment, and then every meal delivery thereafter (what, how many, by whom). We had data sheets where we recorded the moment-by-moment activity of the thirty burrows.

We took great care to ensure that our own presence didn't affect the birds' behaviour. We made our observations from a small pale-green canvas blind, equipped with a bench seat and a window overlooking the puffin slope. We crawled behind a hedge of low, shrubby trees to access the blind, so that the birds would have no idea we were there.

The majority of parental shift changes and food deliveries happened at dawn and dusk. Because puffins nest in burrows, it is impossible to see who is in there. It was paramount that the person with the first shift was in place before first light, ready to see who emerged, and which of the pair had spent the night in the burrow.

At times it was impossible to keep track of their comings and goings, even though we were watching a select few burrows. It was hectic and confusing—and all went to hell when an eagle flew over, clearing the slopes.

Although Libby had already spent a summer on Gull Island, she hadn't adapted to the cold climate that came with the Labrador Current. She wore long underwear to bed, wool socks and a wool cap. She took a banana with her into her sleeping bag, to refuel in the middle of the night. In June.

We slept on bed frames covered with cardboard from disassembled food boxes, to cushion the metal slats and provide a smooth surface. Below the floor of the cabin, in the soft forest loam, there were more burrows—and from them came the nocturnal greetings of the Leach's storm-petrels, a delicate seabird the size of a robin. By day, you would never know they existed. But if you were on the island at night, the sky and the very ground beneath you came alive.

Inside the burrows, the petrels would greet each other with noisy glee, a combination of ecstatic chatter and rising coos, less than a metre from your resting head. Sleep was impossible without the aid of exhaustion and earplugs. The students and researchers who were permitted to stay overnight on the protected islands were very familiar with the racket. I didn't mind. A lifelong insomniac, I enjoyed the company. The cheerful greetings offered me a new perspective on what it meant to be awake at three o'clock in the morning. When the petrel coos softened and finally stopped, it was almost dawn. Time to get up.

We took turns doing the dawn shift. Libby often had work to do in the evenings, so I sometimes volunteered to do extra early shifts. Crawling out of bed in the cool darkness, pulling on the layers of wool, a down jacket, waterproof pants. Hat and gloves. The knapsack already packed the night before with data sheets, pencils, clipboard, binoculars, and the essentials for early morning—a Thermos of coffee, Purity sweet bread and cheese. The sweet bread was as hard as a rock, a baked lump of lard, flour and sugar. It took hours to eat, one of its charms. A headlamp to help you find your way along the short stretch of wooded trail before you veered off on hands and knees, keeping below the low scrub, to the blind. Wrestling the knapsack off too many layers of clothes in a confined space. Pull off the gloves, unpack the clipboard, pencil and binoculars. Pour the first cup of coffee into the Thermos cup. Settle. Sip. Wait.

A puffin lands in a flurry, feet forward like landing gear. The fast wingbeat stops, wings held high for a moment in suspended animation then folded neatly at its sides, collected. The first bird of the day—a quick check of the burrow number and leg band colour combination with the binoculars. Before long, its mate would appear at the burrow entrance. A check of the second bird's legs as the pair greeted each other, nodding their heads, clapping bills, bowing. A moment of puffin tenderness, carefully recorded. And when all the essential data were collected, I could just relax and take in the shared intimacy, how their bond was strengthened and maintained. It felt like a privilege to witness.

As dawn broke, the pace of activity on the slope quickened, pair exchanges happening everywhere, the moment of quiet reflection over. The activity of the early morning shift made the time fly. By 8 a.m. my food was gone, the sun was higher, a few layers shed. Time for a leg stretch. Libby would appear on hands and knees, a few pieces of information exchanged: a band unseen, a chick hatched. And then, like the puffins, we changed shifts.

Afternoons were usually much quieter. There could be lots of birds on the hillside, but fewer exchanges. You had to remain vigilant in case something happened, but a lot of time was spent just watching puffins being puffins—their slow, shuffling Muppet gait, puffed-up chest and colourful bill. The big gape when they yawned. The inexplicable fascination with a buttercup. The unlikely aggression when a stranger came by, a Vise-Grip hold of bill on bill, tumbling down the hill in a feisty tangle. On land they were comical, vulnerable and completely endearing; it never got old.

Weeks stretched into months of eight-hour days in the blind, and we got to know the birds' nuanced behaviours. The hunch of the shoulders and self-conscious approach to the entrance of a burrow—you knew it wasn't theirs. A few tentative steps, disappearing inside (anyone home?), quickly followed by a furious chase out of the burrow, rejected in no uncertain terms. We came to recognize individual study birds, checking the leg bands to confirm. Blue-green/left was conscientious, always arriving on time, always with food. Red/right was a delinquent, appearing for incubation shift and dutifully entering the burrow, only to re-emerge to sun himself in the grass as soon as his partner left. He would enter only for brief periods. Their chick didn't make it; this came as no surprise. You can't tell a puffin's sex by its appearance, but I had a hunch. It is only through a series of body measurements, which Libby had taken, that sex can be determined. My admittedly unfair and completely anthropomorphic assumption proved true: red/right was male.

A few herring gulls nested amidst the puffin colony. These are the large grey-backed gulls with black wing tips that can be found in most fast-food parking lots across North America, eating discarded fries, burgers and chicken bones. They are highly adaptable, able to exploit whatever opportunities present themselves. Some people admire this quality. I am not one of those people. Not on a seabird colony, anyway.

The vast majority of gulls nested together in their own colonies on other parts of the island, but a few had become puffin specialists. They would hang around the entrance of puffin burrows, marauders lying in wait. A puffin returning with fish for its chick would try to fly straight into its burrow to avoid the gull—not an easy feat for a clumsy flyer. Sometimes they succeeded; sometimes the gull would snag the fish out of their bill. Or, missing at their first attempt, the gull would grab the puffin's tail feathers and shake until the puffin dropped the fish and took off again. The chick missed a meal, and the parent headed out to sea with furious wingbeat to the nearest aggregation of fish, to dive in search of more sustenance. All that wasted energy. Meanwhile, the gull stood in the grass, picking up its loot, feeding hunchbacked, nagging chicks. Later in the season, the gulls would snatch naive puffin chicks that wandered too close to the entrance of the burrow to exercise their wings. The gull grabbing the chick headfirst, swallowing it alive. Choking it down took some effort, the gull's neck distended in a grotesque parody of the chick's shape. The little feet flailing helplessly, paddling the air until they disappeared down the throat.

Walking on the island was risky to the delicate surface and disturbed the burrow residents below. The only place to safely hang out was on the roof of the cabin, and that is where I spent most of my solitary time, gazing at Green Island to the south, a busy murre colony that looked like a disturbed hornets' nest from a distance. So many birds on the water. Puffins dipping and splashing, cleaning up after a long stint in the muddy underground. Others holding fish, perhaps enjoying the last moments of calm before returning to the island's demands: hungry chicks, thieving gulls, predatory eagles and hours in a cramped burrow. Socializing. Relatives at a family reunion, mingling but avoiding direct contact. From above you could see the feet paddling underwater, steering them in the direction of choice—puffins a bright orange, murres and razorbills a more stately black. The party was interrupted at times by a surfacing whale; a blow or a massive open mouth, scooping up fish it had driven to the surface. Gulls swooping in to snatch fish out of those mouths, careful to exit just in time. Towering icebergs melted and rolled, the occasional explosion as they broke apart. Fishing nets in neat lines, with colourful markers, up to the island's shore. The wonder of it.

———

When I returned to Gull Island in 2008, it was in the position of Ecological Reserve Manager. My job was to oversee all activities on the islands and in the marine portion of the reserve—organizing tour boat workshops, ensuring regulations were being adhered to, trouble-shooting wherever necessary, and assisting with research and monitoring. I spent a great deal of time that first summer on tour boats or conducting boat-based surveys with the Canadian Wildlife Service. But I had retained the conservation ethic I had been schooled in when I worked there as a field assistant in the mid-eighties. I wouldn't set foot on the sensitive burrow-laden island unless it was specifically required. Ditto for spending the night.

The need to visit the island did not present itself until late in the season, when the presence of a mink was reported to me. They are capable of swimming the kilometre distance from the mainland shore and sometimes do, although we hadn't experienced it in 1985. Mink are not native to Newfoundland, but a strong population has been established and spread throughout the province from mink farm escapes. They are indiscriminate killers; not only will they hunt to eat but, like cats, they will kill just for the sake of it. Mink can wreak havoc on a seabird colony. A graduate student had captured video footage of a mink in one of her puffin study burrows—the silhouette of whiskers and an ear, a glimpse of a startled chick before the mink obscured the view with its body. A shrill cry, clipped short.

I bought live traps and spoke to local wildlife experts about where best to set and how to bait them.

Mink like to travel along streams, so I brought one of the traps to Petrel Creek, hoping to capture the animal while it was on the move at night. I disguised the trap with vegetation and baited it with sardines. On an island covered in burrows with chicks and dropped capelin, the chances of luring it to my trap were pretty slim, but perhaps still possible. I wasn't ready for bed, so I sat in a lawn chair outside the cabin door and waited for it to get dark, for the petrels to fill the air with their haphazard flight and for the nocturnal serenade I remembered from my days on the island twenty-three years before.

There were petrels, all right. They flitted past at fairly regular intervals. If you hadn't experienced Gull Island at night twenty years prior, you might even be convinced that there were lots. Maybe it always started this way and I had just forgotten? So I waited a few more hours, dread filling me like a slow pour of concrete. That night I wore earplugs to bed, to block out the silence.

For Leach's storm-petrel—a species with a huge population that live in burrows and are active at night—changes in numbers are difficult to detect and the population had not been assessed in the intervening years. But the decline was undeniable and stark.

This was the first time I had personally observed a drastic population decline. I had worked on a failing black-legged kittiwake colony in Alaska, and so few chicks survived to fledging that our study season was cut short. But I had no experience with that colony in better times, and kittiwake colonies in my part of the world were flourishing. The personal impact was minimal. But what had happened to the storm-petrels on Gull Island was crushing.

———

Species differ in terms of their vulnerability to risk, depending on their life history. Leach's storm-petrels do not dive or spend a great deal of time sitting on the surface of the water, and are not as vulnerable as the auks to fishing gear entanglement and oil spills. But their adaptations for a nocturnal lifestyle, a strategy that evolved to keep them safe, may contribute to their undoing.

Storm-petrels feed on tiny marine fish and crustaceans that are bioluminescent—they have small organs called photophores that light up in the dark. These small prey migrate deep into the water column to protect themselves from surface-feeding birds and mammals during the day. At night, they migrate to the surface, emitting a sparkling glow into the water—unsuspecting targets for these voracious little birds. Light means food for a storm-petrel, and their attraction to it is hard-wired. This adaptation has served them well and the species has flourished in the North Atlantic, particularly around the rich waters of Newfoundland.

A single foraging trip for a parent lasts around five days and can take it as much as a thousand kilometres from the breeding colony in search of food for its chick. Most go to the Grand Banks and the adjacent shelf waters, where upwellings create areas of high productivity. A four-hundred-kilometre one-way trip as the crow flies—but petrels don't fly like crows. In the absence of strong winds, their flight is light, jaunty and unhurried, with no loyalty to any particular course. When hunting, they fly in a zigzag pattern over the surface, changing directions on a dime, a pattern that suggests a child on a sugar high. They pick small creatures off the water, with their wings raised, their feet gently paddling, barely touching the surface. This feeding habit has earned them the nickname "Jesus bird" among fishermen who have witnessed it, because they look as though they are walking on water.

The adaptations for light attraction evolved for a dark, expansive, open ocean. The problem is, the ocean isn't so dark anymore. There are thousands of ships on its surface and all emit artificial light at night. Storm-petrels are attracted to the light and they will follow ships, circling in a trance. Disoriented, they may collide, causing injury or death. Or they will land on the ship and, on apparent solid ground, seek darkness; the safety of a burrow. Hiding under pipes, stairs, winches—anywhere that gives cover. In the process, they may encounter lubricants, hydraulic fluid, cleaners—any number of toxic chemicals used in the routine maintenance and running of a ship. Under certain wind and weather conditions, hundreds can be drawn to a single ship, the crew having to shovel them off the decks and into the water.

Offshore oil platforms are permanent structures that act like artificial reefs and attract seabirds, including storm-petrels. Like ships, they are well-lit, but they feature one extra lethal element that ships don't have: gas flaring. In the oil and gas recovery process, excess gas is burned off, creating a large flame that glows twenty-four hours a day, lighting up the night sky. At the edge of the Grand Banks, where storm-petrels forage to feed their young, there are currently three platforms. What does this mean for a species that is attracted to light like a moth to a flame?

There is no question that offshore oil production threatens adult storm-petrel survival. But it is not the only threat. Since the cod moratorium, the reduced availability of fish offal from fishing vessels and processing plants has left gulls looking elsewhere. The island paths are sometimes littered with petrel wings—evidence of a rough night for adult storm-petrels. Maybe a moonlit night had offered gulls better opportunities to hunt them before they could reach the safety of the burrow. It is possible that less obvious threats are also at play.

I left Gull Island after that second visit deeply worried about the state of the storm-petrel colony. Leach's storm-petrel population surveys had not been conducted in decades, when the population for the Witless Bay colonies was estimated at 700,000 pairs. Most population management efforts were focused on murres, a legally hunted species in Newfoundland and Labrador that requires close monitoring. In 2011, Dr. Sabina Wilhelm, a seabird biologist with the Canadian Wildlife Service, started preliminary work in Witless Bay and Baccalieu in preparation for the immense task of updating the long-overdue population estimates. In my role as seabird reserve manager, I helped her with it.

Sabina and I started on Great Island, which is also part of the Witless Bay Ecological Reserve but is slightly more remote than Gull Island and much more dangerous. The landing site is located in a gulch where, even on the calmest days, water funnels in and churns up, ramming the walls from all directions before being sucked out again. At the end of the gulch there is a cave ominously known as the Devil's Throat.

Careful calculation of the wind and wave conditions is always part of any seabird colony logistics plan, but particularly for landings like this. The research community has a healthy respect for the challenges of accessing Great Island, and it is visited infrequently by students and scientists alike. Still, there is a very small research cabin on the island. We picked a day with a good sea state and headed out. A tricky combination of ropes and rock-ledge footing and we were ashore.

For species that nest in burrows, population censusing is particularly difficult. It is not a simple matter of counting visible birds sitting on nests. The process is much more labour-intensive and requires sticking your arm deep into the earth, down the length of the burrow, to discover what is at the end—an adult, an egg, a chick. Or, if it is inactive, nothing at all. This is called burrow "grubbing."

Petrel burrows are made to fit petrel specifications. They have tiny entrances, barely large enough for a small hand to fit through, and are often very deep, much longer than an arm, with sharp turns that bend in a way the human arm does not. The burrow is frequently intersected by roots of trees and ferns. Grubbing to the end of the burrow is often difficult and sometimes impossible. Sabina was experimenting with recordings of petrel calls. The idea was to play the recording at the entrance to the burrow and see if it elicited a call back from the inhabitant and how reliable this method might be for determining burrow activity without having to grub. All burrows where recordings were played were to be grubbed, regardless of apparent activity. The results of the call-back prediction of burrow occupancy would then be compared to the results of hands-on grubbing.

The study was carried out at night, since storm-petrels are naturally vocal in the burrow then. In daylight, we set up the boundaries of the study plot we would sample. At midnight we headed out into the pitch-black, blinded to the surroundings by the bright, narrow beam of light from our headlamps aimed a few paces ahead. We walked in silence, with an occasional warning. *Careful!*

At each study plot we split up, walking to opposite ends. The burrows are too narrow for sleeves, so we took off our thick oilskin jackets and placed them on the ground to lie on. We then retrieved our sampling equipment—a playback recorder, Popsicle sticks to mark the burrows we had sampled and notebooks to record results. Lying on the ground with the recorder aimed at the burrow entrance, we hit Play and listened closely. Sometimes a responding chatter could be heard, but it wasn't always clear which burrow it was coming from. The next step was to grub—lying with your cheek pressed into the earth, arm stretched down the tunnel as far as you could possibly reach. Trying to navigate the turns. If the turn was in the opposite direction of your natural elbow bend, that meant rolling over and trying the other arm, scraping against rocks and tree roots until you finally reached the end of the chamber. Then a fingertip assessment of its contents—the soft bite of an adult, the fluff of a downy chick—before retreating, carefully repairing any minor damage and moving on.

I had not forgotten the possibility of mink. My trapping attempts in 2008 had failed, but that didn't mean they weren't still there. And if they'd made it to Gull Island, it was possible they had reached Great Island as well. Would a mink fit in a storm-petrel burrow? Did I want to find out with my hand? Each time I committed my arm to the blind task, I thought of this. There was an undeniable uneasiness that accompanied the whole business. But each time I did not draw back a bloody stump, my fear of a mink encounter lessened. Sharing in the underground intimacy of the night, I imagined the work it took to dig and shovel the long tunnels, with webbed feet no larger than my thumbnail. The trips back and forth. Hitting obstacles; changing course. And also, returning from a thousand-kilometre journey at sea to this exact burrow entrance amongst thousands of others, and to the familiar and recognizable sound of its mate, deep within.

In the end, the call-backs didn't prove to be reliable enough to replace grubbing. Still, it was worth a try at the onset of an intensive Leach's storm-petrel breeding population assessment, which would encompass all the major colonies, including Baccalieu Island, the largest Leach's storm-petrel colony in the world.

When talking about seabird populations at a breeding colony, the numbers are usually presented as "breeding pairs." Each pair has a chick, and non-breeders will hang around a colony prospecting or socializing before maturing to breeding age. The number of chicks that survive may vary wildly from year to year, and the non-breeders may also come and go throughout the season. The breeding pairs are the bread and butter of a stable colony.

The population census confirmed what the night sky had already revealed: Leach's storm-petrel numbers were way down. The Baccalieu Island Reserve, more remote than Witless Bay and relatively difficult to get to, hadn't been surveyed since the mid-1980s, when populations were estimated at about five million pairs. The results of the updated surveys were alarming to say the least: Witless Bay was down from 700,000 to just over 300,000; Baccalieu Island was down to about two million pairs. Over 3.5 million birds were unaccounted for; both colonies had lost over half their population.

These two colonies, along with a colony on the adjacent French islands of Saint-Pierre and Miquelon at Grand Colombier (about 364,000 pairs), make up close to half the world's population. Leach's storm-petrels travel immense distances to their wintering grounds. Research on the Baccalieu Island and Witless Bay birds showed that they will spend the winter months as far away as the east coast of Brazil and the southwest coast of Africa. What risks could they be exposed to during the vast ocean migrations?

That same summer of 2011, seabird researcher Alex Bond was investigating plastic ingestion by storm-petrels. He captured storm-petrels using mist nets—a bird version of a gillnet, set in the air. Only birds are not killed; they are removed from the net and "processed"— data are collected from them and they are subsequently released. This usually includes taking various measurements; banding; sometimes taking blood or feather samples; or attaching a tracking device. In this case, Alex was collecting stomach contents by plumbing them with fluids, forcing the birds to regurgitate. He was specifically interested in analysing how much plastic was recovered from the stomach samples. Unfortunately, the method was deemed too stressful for the birds and the study was discontinued. But it was clear from the few samples he took that parents were feeding their young chicks meals laced with plastic.

Is plastic an important contributor to the decline in petrel populations? Alex didn't think so. Though its impact will surely increase in response to the continuous flow of plastic in the ocean, there are also many other potential culprits. The closure of the cod fishery led to the closure of fish plants too. The ready and plentiful supply of fish offal provided by the plants was suddenly cut off and gulls were looking to other sources for food. Were they hitting petrels harder than ever? There is also evidence that petrel diets are shifting, likely in response to changes in availability of their favourite prey, the tiny bioluminescent fish known as myctophids. This of course leads to the question: What is happening to myctophids? Are they part of the collateral damage imposed by seismic blasts in the petrels' primary feeding area, where offshore oil development and production is happening? Or are they affected by changes in ocean temperatures and currents? Is the increase in offshore oil production platforms, supply vessels and oil carriers—and the attendant lights and flaring—simply multiplying the number of petrels that die there each summer while attempting to secure food for themselves and their young?

The answer is no doubt a complex combination of all the above. While scientists are working hard to tease out the strands, it feels like a race against time. In 2022, the deep-water Bay du Nord oil project was approved for development in Newfoundland's offshore—on the slope of the Grand Banks, where petrels zigzag toward danger in search of food.

Chapter 5

Full Throttle on Baccalieu Island

Cape St. Mary's was the only northern gannet colony I knew until I became the seabird reserve manager of Baccalieu Island, off the northwest tip of Newfoundland's Avalon Peninsula. For an avian biologist, Cape St. Mary's is irresistible, a riot against the senses. The racket, the smell, the birds' aggressive assaults on neighbours, followed by gentle preening of mate and chick—all rituals that indicate a thriving colony. But over the course of many years I have noticed an increase in the fishing net and rope that have been incorporated with grass and seaweed into their nests. While it is true that the moratorium on the Atlantic cod fishery was a game-changer for the auks, who forage close to their colonies, gannets can travel hundreds of kilometres in a single foraging trip, expanding their potential interaction with fishing gear. The gear from the other fisheries, along with ghost gear, continues to provide a threat from bycatch in the water—and on land, in the form of nesting material. I was keen to see how the colony on Baccalieu had held up in comparison with the more accessible southern seabird reserve. It would become clear that ghost gear and plastic debris, though present in nests at both colonies, are not the only threats to the gannet populations.

More remote and logistically difficult to access than either Witless Bay or Cape St. Mary's, Baccalieu hadn't received much research attention and populations hadn't been monitored in decades. My first priority was to get a sense of the island and the distribution and abundance of its seabird inhabitants. I arranged to circumnavigate the island with a fisherman from nearby Red Head Cove in the summer of 2010. Although the community was named for the iron-rich headlands of the cove, it could easily have been named for its residents. With a head of rusty hair, Gary was no exception. We met in Bay de Verde where he kept his boat and then headed to Baccalieu to get an overall picture of the seabird activity on the island from the water.

It was calm and sunny. A trip around the island, close enough to the cliffs to observe nesting seabirds, called for ideal conditions. The surrounding wall of towering cliffs was spectacular on a still day, but I couldn't help thinking how forbidding it would be in a storm. The early morning light was swallowed in shadow as we approached the island from the Tickle—the narrow channel that runs between the island and the mainland—on the west side. Gary slowed the boat and the engine quieted to a gentle putt-putt. From the dark cliff face, the barely there high-pitched notes of the black guillemots. Emerging and disappearing into the shadows of the rock face, the whirring wingbeat, the strobing flash of white on the wing. Almost hallucinatory, if not for the solitary birds strung out in a line along the contour of a narrow crevice, like knots in a rope. Proving their existence.

The name *Baccalieu* is derived from the Portuguese *bacalhau*, meaning cod, and speaks to the historical importance of the island to the cod fishery. Like all the major seabird colonies in Newfoundland, it is located on the east coast, where the greatest force of the Labrador Current exerts itself. Five kilometres long and one kilometre wide, it runs adjacent to the cliffs of mainland Newfoundland, separated by a narrow channel (the Baccalieu Tickle). The island, which sticks out past the tip of Conception Bay, is exposed to the open North Atlantic, often buffeted by gales and shrouded in dense fog. On clear days, it is rich and spectacular and completely unforgiving.

There are only two places to land safely on the island, if you use the term *safe* loosely—Lunin Cove on the northwest side of the island and Ned Walsh's Cove on the southeast. The latter was named for the only fisherman to settle full-time on the island. In the late 1800s, he and his wife Ellen had a vegetable garden, kept animals and raised eight children there. A seasonal fishing community operated out of Lunin Cove, and when a load of salt fish was dried and ready to be shipped, a flag would be raised from the highest point, visible to the nearby community of Red Head Cove. Once this signal was received, a schooner was dispatched to collect and bring the fish back to the merchant's premises.

Every nook and cranny of the island's coast has a name, a story telling of the history and treachery of the place: Devil's Under-Jaw, Mad Goat Gulch, Cabbage Garden Point, Linda's Rocks, Lassie Gulch, Cow Path Head, Murdering Hole. And the seabirds, with their local names, are as much a part of the island as the cliffs themselves—Tinker Hole, Gannie Cliff, Pidgeon's Point.

In the past, Baccalieu had two lighthouses—one at the north end and one at the south—and two lighthouse keepers for each of them, working opposite shifts. The island's terrain was covered by deep valleys, bogs and impenetrable tuckamore—thick, stunted spruce that grow as high as they dare, sheared off against the direction of the prevailing wind. The whole grove leaning in as if to hear a final verdict, a testament to the wind's relentless force. Despite the challenging terrain, the lightkeepers made a trail between their houses, taking advantage of high, treeless ground where possible and carving a path through the tuckamore when necessary. It was an arduous ten-kilometre round trip, but worth it for the occasional company.

The island was protected as a seabird reserve in 1991, with special permission granted to allow maintenance activities there. Around this time the north end lighthouse was decommissioned, replaced by an automated beacon on a lattice tower. The old tower was left standing, a vestige of a bygone era. At the south end lighthouse, the precarious old system of ladders and pulleys that had been used to land goods and material from boats to the lighthouse was left to crumble, replaced by a helicopter pad. In 2002 the doors to the house were closed for good. There remain subtle traces of the old family settlement in Ned Walsh's Cove—the remnants of gardens planted long ago, now used as protective cover for Leach's storm-petrel burrows.

As in Witless Bay, the auk populations here had increased dramatically, benefiting from the cod moratorium and the resulting lack of fishing gear in the water. In the eighties, the Atlantic puffin colony had not extended into the grassy terrain of the two landing areas at Lunin Cove and Ned Walsh's Cove, perhaps because of their history of human habitation and the tradition of berry picking that had persisted long after. In the absence of disturbance, coupled with a need for more puffin burrow real estate to accommodate the increasing population, puffins had expanded into the grassy slopes at both locations. Razorbills, barely a footnote decades ago, were in groups of hundreds on the water. At Northern End Gulch, seven bald eagles hunted over the kittiwake colony, a shot of panic ricocheting off the cliffs, sending a spray of birds into the air. Eagles swooped in to snatch an exposed chick, momentarily abandoned in the confusion. The constant nasal scream of the kittiwakes, magnified by the surrounding rock walls. Life and death at full throttle.

The east side of the island was drenched in sun by the time we got there. The entire coast busy with activity, dominated by whichever species was offered the best option: a grassy slope (puffins), a cliff face (kittiwakes and murres), a boulder bed (razorbills). And then, at the southeast end of the island, Gannie Cliff, where over two thousand pairs of northern gannets nested in a symmetrical grid of bulky nests on two-metre-by-two-metre territories, the boundaries fiercely defended. The gannet's bill a filed saber, perfect for jabbing and jousting with any neighbour foolhardy enough to stretch a wing or extend a leg beyond the invisible border.

There were hundreds of thousands of breeding seabirds in evidence everywhere along more than ten kilometres of coastline. And resting in their burrows were the world's largest Leach's storm-petrel colony, silent and invisible. The island pulsed with the heartbeat of over six million birds.

Later that summer of 2010, I had the opportunity to land on the island with my master's supervisor and long-time friend Bill Montevecchi. Bill had conducted the last estimates of Baccalieu's bird populations back in 1985. In 2010, he and a German colleague were collaborating on a study that involvedtracking gannet movements to see where they travelled on foraging trips to feed their young, and where they spent the rest of the year away from the breeding colonies.

Handling wild birds requires finesse, no matter their size. You have to know how the bird is built and what it can take. For small songbirds, the wings are very flexible and have a generous range of motion, but if you try to straighten the toes of a bird with a bent leg, the toes will break. In the case of a gannet, you have to apply your strength but know when to be careful and gentle. Some observation and instruction on gannet handling protocols was essential.

Gannet catching is a mucky, filthy business. It requires a great deal of stealth and patience. And I had seen what gannets do to each other with those bills—I did not want to be on the receiving end of that wrath. I was assigned the benign task of recording the data, a job I happily accepted. But like all good mentors, Bill encouraged me to change roles once I'd seen how it was done.

Gannets are captured using a noose made of monofilament fishing line that hangs from the end of a telescopic pole. I'd had some experience with this equipment, catching murres on other islands. But gannets are much more wary than murres, and there isn't the option of appearing from nowhere on the cliff edge above their nests as you can with murres, which takes away the element of surprise.

I wore a hat to obscure my face from the savvy adults and crawled out slowly on my hands and knees, looking away from the birds. I settled in and let the birds acclimatize to my inanimate presence, a change in their surroundings. They eventually lost interest. Then I extended the pole slowly along the ground, the noose wide open, facing up. When it reached the targeted bird, I raised the noose slowly from the ground, in front of its chest and over the head, with a quick jerk to tighten the line and walk-drag the bird away from the others. This is where a bit of courage and pluck is required—approaching the gannet and getting control of its neck before it can stab you, covering its head in a brin bag. Glasses and leather gloves offered a fallback in case the bird was quicker than me. But I really didn't want to test the performance of my flimsy equipment against this force of nature.

Once at the processing set-up, I kept the bird immobile while Bill banded, took blood and attached the geo-logger. Above all else, I needed to hold that bill steady and closed. This required force against brute strength, but there was also a fragility there. Gannet nostrils are flush with the bill and hidden along a black groove—easy to miss. The deal: for the next few minutes you will be stressed and defenceless, but I will be careful to protect you from harm.

Processing was done as quickly and efficiently as possible, the conversation like that of surgeons in an operating theatre: *Calipers. Band. He's breathing heavily, watch the nostrils. Is the needle site still bleeding? No? Okay, let him go.* Then, turning the head toward the open ocean, a quick yank to remove the bag and—take off! The reassuring appearance of the bird back on the nest a few minutes later, preening to reorganize the feathers that had been ruffled in the ordeal, then picking up where it had left off.

The geo-logger data collected from the Baccalieu gannets were part of a larger study involving all the major gannet colonies in North America. Surprisingly, all these colonies had individuals that travelled not along the east coast of North America, like the vast majority, but across the Atlantic to Europe. There they joined the migration of gannets from European colonies all the way to the west coast of North Africa. Why this route, so much farther to travel? A route with ship traffic, ghost nets and other floating plastic. Was it safer than travelling the populated and polluted eastern seaboard of North America? And what of the European coastline? Or was it more likely a primal instinct: birds physiologically wired to return to a colony of some ancient relative, an instruction somehow imprinted in their own DNA to return home? Do they take advantage of the prevailing winds to carry them across the ocean? And what will happen if these winds change, as climate change alters the course of wind and sea?

———

Gannet capture for this work was best conducted in September, late in the season, when chicks were large enough to fend for themselves for the brief period that the parent went missing. Bill and I returned to the island several more times over the following seasons, both to deploy more geo-loggers and to recapture birds to retrieve the precious data that had been collecting for a year in the bands strapped to their legs. On one of these visits Bill left early and I stayed behind to finish the work with one of his graduate students. Paul and I spent hours each day bent to the task of catching gannets. One day, near the end of the project, we heard the forecast calling for strong winds. This wasn't a shock—fall is a windy time of year. We figured we could finish deploying the remaining loggers before the wind hit, and if not, we would just wait it out.

New birds are relatively easy to catch; they have no experience with the noose and all that it entails. But generally speaking, you have only one chance. If you miss your mark with the noose and the bird manages to escape, you rarely get a second opportunity. Once wary, they are almost impossible to catch. For the same reason, recaptures, even a full year after initial capture, are incredibly difficult.

We took turns crawling through the mud to a position that gave access to the target bird, if all went well. You extended the pole in barely perceptible increments past other nests, holding your breath when a chick occasionally stepped into the noose. Then out again. Breathe and keep going, inch by inch. The quiet, patient approach, suddenly pierced with panic (the bird's) and speed (yours). The quick but careful retrieval of the bird, ensuring both the bird's safety and your own. Then you folded the five-foot wingspan by its sides and carried the three-kilogram bird under your arm to the processing station. The bird fighting for its life the whole way.

I've always liked the feistiness of seabirds when you capture them—the twisting and clawing. It is a reassurance that they are fully alive. But this seemed a bit much.

Between captures, Paul and I hid out of sight and allowed the colony a rest period, a chance to get back to normal. By now the chicks were large, confident and naive—apparently not overly concerned by the disturbance—and easily found their way back to their nests. But we did notice one small downy chick that wandered around, looking lost and getting poked, despite its diminutive size, by disgruntled neighbours.

While the colony regrouped, we spent our time taking photos, chatting quietly, setting up supplies for the next round.

The work went well and we finished up just ahead of the strong winds. Perfect timing—we would use the stormy conditions to organize and pack, ready for the helicopter the following day.

For the first few hours the storm was a novelty. We challenged our strength, spreading our coats and leaning into the wind, balancing there. We took pictures of the waves as the sea built. But as the wind force continued to increase, we retreated to the safety of the light-keeper's house to ride it out. We had no radio contact and no idea what was being reported, but we could tell this was huge. The winch building perched at the top of the cliff disappeared from view with each crashing wave. Spray started to hit the windows on the cliff-facing side of the house, thirty metres above the sea. We were afraid the windows might shatter, sending a spray of glass and water across the room, and nailed two-by-fours that we found in the basement across them to offer some protection.

The house construction was solid, but we had no idea how much its structure had been compromised by almost a decade of neglect. From the couch, I noticed the west wall bowing inward slightly. A Beaufort scale force 8 raged in the toilet bowl. Somehow, the wind was blowing up the plumbing, sending sprays of putrid water into the air. Closing the lid wasn't enough; we shut the bathroom door and didn't open it again.

The winds subsided during the night, giving way to a spectacularly sunny morning. There was little evidence of the storm that had raged the day before, other than the telltale roof shingles strewn across the ground—breaches in the roof's waterproofing that would lead to its undoing. But for now, the building was fine.

We had no idea that a full-blown hurricane had ravaged eastern Newfoundland, that trees were upended, power lines down, highways flooded and washed out, communities completely cut off. A state of emergency had been declared and the support of the military had been engaged. All helicopters in the province were on standby for emergency relief. We didn't know it, but we were stranded.

How exactly does a seabird colony react to such an intense weather force? We decided to hike over to the gannet colony to see how they had fared through the storm. To our surprise, there was no discernible change; at first glance, everything was exactly as we had left it.

But I was curious about that one chick, still downy and too young to survive that late in the season. We looked for it, but it was nowhere to be seen. Until Paul noticed the outline of the young bird on the ground, completely flattened, wings extended and covered in mud, like a fossil. Evidence of the pandemonium that must have occurred—the chick crushed by a panicked stampede, the strict enforcement of nest territories temporarily abandoned. But for now at least, order was restored.

———

Gannets are tough and resilient. Like all seabirds, their survival is calibrated to a specific set of conditions—their arrival at the colony, courtship, egg laying and chick rearing timed with the tried-and-true promise of plenty of food to feed and raise their chicks. The colony location is chosen for proximity to a rich food source during the breeding season, when they need it most. But the unusual storms are becoming more commonplace. Historically, tropical hurricanes lose their strength when they hit colder bodies of water, so that by the time they reach the Labrador Current they usually fizzle out to a post-tropical storm. But with the warming of the oceans, storm events are occurring earlier and more frequently, and are retaining much greater force. The frequency of tropical storms and hurricanes in Newfoundland has more than doubled in the last century.

In 2013 the provincial government—the jurisdiction responsible for protecting the islands—laid off all its seabird reserve managers. I was one of them. It was a cost-saving measure that removed any meaningful oversight of activity in the reserves, as well as the ability to monitor conditions on the ground. There is no one now to watch over the place.

Hurricane Larry hit eastern Newfoundland in September 2021. High winds and storm surge wiped out part of the colony on Bird Rock at Cape St. Mary's and blew gannets away from it, some injured and stranded kilometres inland. How the gannets on Baccalieu fared during and after that hurricane, unfortunately isn't known.

As well as changes in storm activity, warming oceans are also affecting the rhythms and patterns of marine ecosystems, with some devastating effects. A marine heat wave in the North Pacific, known ominously as the "Blob," formed in the Gulf of Alaska in 2013, spread south to Mexico's Baja peninsula and lasted until 2018. Its impacts were devastating. Marine ecosystems were turned on their head; a hundred million Pacific cod disappeared, whales didn't show up to their usual summer feeding grounds, and many were found dead on beaches. A million murres alone were estimated to have died as a result of the warm water event. Mass die-offs have become increasingly common throughout the world's oceans; climate change has been identified as the main reason for the decline of sixty-three seabird species.

A warming of the North Atlantic in 2012 proved devastating for northern gannet breeding success that season. Around the week of August 12, 2012, an unprecedented mass abandonment of chicks by their parents happened concurrently in at least three of the major gannet colonies in North America—Cape St. Mary's, Great Bird Rock on the Magdalen Islands, and at Île Bonaventure at the tip of the Gaspé Peninsula in Quebec, the largest North American colony. In the absence of oversight, it is not known whether the phenomenon occurred on Baccalieu. The abandonment is thought to have been precipitated by the warmer than normal water temperatures that made conditions untenable for mackerel and caused them to either move farther north or retreat to cooler, deeper waters—out of reach of the gannets. Either way, the gannets were driven by the threat of starvation to abandon their nests, some for over a week. Few chicks survived. Though not as severely, this never-before-seen abandonment phenomenon occurred again in 2014, 2015 and 2018.

The summer breeding success of 2022 has once again proven a failure for gannets and other seabirds in the North Atlantic. An avian flu outbreak ravaged seabird colonies, but it is probably not the only culprit. On the east coast of Newfoundland, where gannets died in droves, there were unprecedented signs of a much warmer ocean. Hundreds of Portuguese man-of-war jellyfish washed up on beaches along the east coast, and a birder's bonanza of Cory's shearwaters were seen near shore in early September. Both species belong to the warm waters of the Gulf Stream, not the Labrador Current. Trying to guess the true magnitude of what has been lost in 2022 and what is likely to transpire in 2023 is like trying to read tea leaves (or predict the trajectory of COVID-19).

Gannets are resilient, all right, but they are not invincible. Human-induced climate change—driven in no small part by plastic production—is creating novel, intense weather systems that they are having trouble coping with on land. And at sea, like all marine life, they are struggling to adapt to changing temperatures and ocean currents, and an environment booby-trapped with lethal plastic debris. Fishing gear, with its plastic floats and nylon ropes and nets—a plastic double whammy.

Chapter 6

Ghost Harvest: The Threat of Macroplastics

We know more about the surface of the moon than the depths of the ocean. Often, science relies on inference to fill in the gaps. Before modern technologies, researchers knew that murres dove to three hundred metres because they were found dead in gillnets that were set at that depth. What happens kilometres beneath the surface is even more mysterious. We glean clues from what washes up on coastlines, and in this era of digital connectedness, news outlets share discoveries with the world in real time.

In February 2017, media reported that a Cuvier's beaked whale had been found starving on a beach in Bergen, Norway. The family of beaked whales to which the Cuvier's belongs is the most elusive of all the whales. They live in very deep waters and come to the surface just briefly and inconspicuously for a few breaths before disappearing into the abyss once again. Some species of beaked whales are only known from corpses that have washed ashore, and the volume of plastic they have been exposed to is only known through the burden they carry. The Bergen whale was euthanized when it became clear that it was dying. An autopsy revealed the whale had thirty plastic bags in its stomach and no trace of food.

Another Cuvier's beaked whale was reported dying off the coast in the Davao Gulf in the Philippines in March 2019. The details are gruesome: The young whale was seen near shore with blood pouring out of its mouth, its body listing as it swam. The resulting autopsy found forty kilograms of plastic in its stomach. The plastic—made up of bags and nylon rope—was compacted to the density of a base-ball, only much larger, and some of it had started to calcify. Stomach acids were useless against this synthetic imposter, and instead burned holes through the whale's stomach lining, leaving it emaci-ated and dehydrated. The following month a dead pregnant sperm whale was found on a beach near a resort town on Sardinia, Italy. Twenty-three kilos of plastic filled more than two-thirds of her belly.

Similar reported cases have become too numerous to keep track of, and of course most cases go undetected and unreported, far out at sea. UNESCO estimates that 100,000 whales and other marine mammals die from plastic ingestion every year. The tragic deaths of whales tend to attract media attention, but the vast majority of less "charismatic" species die in droves, unreported.

The impact plastics have on marine life are many and varied, depending on the type, the size and the location in which the plastic finds itself. Floating bags or sheets of plastic will drift through the water column and gradually sink. They may weather and break apart along the way, be mistaken for food and eaten, or land at the bottom intact, smothering life on the ocean floor. Most likely they will meet some combination of these fates. Any plastic greater than five millimetres in size is considered a macroplastic; inevitably, it all becomes microplastic that is invisible to the naked eye but exists in the ocean as a toxic soup—a plastic smog.

By and large, the greatest threats plastic poses to marine wildlife are from ingesting it or becoming entangled in it. These threats are becoming increasingly significant as the worldwide concentration of ocean plastic increases.

When it comes to ingestion, some marine predators mistake plastic for prey and actively pursue it. Species that feed on soft-bodied animals are vulnerable here. Sea turtles, toothed whales and many seabirds feed on gelatinous prey like squid and jellyfish and are genetically hard-wired to pursue them. Soft plastic mimics the shape and movement of these prey as it undulates in the water column. The soft-bodied prey and the animals that feed on them are often located far out at sea or at depths well below the surface, unseen. But that is exactly what I witnessed off New York City on Valentine's Day in 2017, with the loggerhead turtle lumbering after that Mylar balloon, endowed with sentimental script and metres of ribbon. Most of the evidence of fatal plastic ingestion is revealed through autopsies, when dead or dying animals wash up on shore, as with the distressing examples of the Cuvier's beaked whales.

On YouTube, you can find a video of a turtle with a plastic straw embedded in its nostril. The video runs for eight minutes, while a group of sea turtle researchers attempt to remove it, grabbing and tugging with pliers. The turtle's mouth gapes from the stress—bleeding from the effort applied to remove the straw, adhered to the delicate membranes within. Finally, whatever holds the straw in place gives way and with a gush of blood the length of straw comes out, bent and twisted but otherwise intact. The video went viral and the straw-in-turtle-nostril image became the poster child for the movement to ban the use of plastic straws. But this was a relatively random event. Not to say that it doesn't happen regularly, and maybe even frequently, given the volume of single-use straws in the environment. Americans alone use 500 million plastic straws every day—enough to encircle the globe two and a half times. But turtles aren't specifically vulnerable to plastic straws. Unlike the undulating plastic in the water column, which turtles will actively—and successfully—hunt. When turtles eat plastic, not only does their system become blocked, but gas builds in their digestive tract. When this happens, the turtles are like inflated balloons themselves; too buoyant to dive, they are unable to pursue prey or evade predators at the surface.

All seabirds are vulnerable to ingesting marine plastics, and the risk is increasing around the globe. But one group, the Procellaridae family, are especially so. These species are also known as tubenoses, because they have an extra nostril tube that runs along the top on the upper bill. There are over a hundred species of tubenoses, which roam the world's oceans and range in size from the small storm-petrels, about the size of a starling—to the wandering albatross, with its three-and-a-half-metre wingspan. These seabirds are known to have a keenly developed sense of smell. Many species in this group feed on soft-bodied prey, and mistake soft plastics for food. For some tubenoses, krill and other small prey form an important part of their diet. Their highly developed sense of smell can lead them to a buffet, quite literally, of junk food. This is how.

Hard plastic forms an ideal substrate for marine algae to grow on. When krill and other zooplankton graze on the algae, it releases a chemical called dimethyl sulphide (DMS), which has a strong scent like rotten eggs. Tubenoses are naturally attracted to the scent, which they associate with krill aggregations. They will ingest the plastic substrate either incidentally, or mistaking it for food, in pursuit of krill.

There are alarming examples of plastic ingestion by tubenoses from around the globe. Eighty percent of the world's flesh-footed shearwater breed in waters around Australia. Long-line fishing, nesting habitat destruction, and predation from introduced rats are known risks to shearwater survival, but plastic ingestion has become an issue of increasing concern. Skeletal remains found with heavy plastic loads in the gut prompted investigation on Lord Howe Island, a small, remote island in the Tasman Sea, between southeast Australia and New Zealand. In a 2005 study, 579 pieces of plastic were found in the remains of fourteen dead birds. The stomach contents of fifty-six living chicks that were near fledging age (ready to leave the nest and fend for themselves) were examined. Forty-four of them (79 percent) contained plastic. The Agreement on the Conservation of Albatrosses and Petrels (ACAP) reports that about 690,000 pieces of plastic are deposited on Lord Howe Island alone each year—either through routine regurgitation of dietary waste (known as a bolus) or from the carcasses of dead birds.

The Midway Atoll National Wildlife Refuge of Hawaii is home to the largest colony of Laysan albatross in the world. It is also close to the Great Pacific Garbage Patch—a 1.6-million-square-kilometre collection of marine debris in the North Pacific Ocean, roughly twice the size of Texas. An estimated twenty-three tons of debris accumulates on the atoll every year. And each breeding season, a staggering five tons of this plastic is fed to chicks.

In the North Atlantic, data on seabird plastic consumption were summarized and reported by Circular Ocean for northern Europe, Scandinavia, Russia, Greenland, Svalbard, the Faroe Islands and

Iceland. Of the thirty-four species investigated in these wild and remote areas, 74 percent had ingested plastic.

Although the tubenoses are probably the most vulnerable group when it comes to eating plastic, the problem is widespread. Plastic consumption by seabirds has been tracked for over sixty years. In 1960, plastic was found in less than 5 percent of seabirds. By 1980, it had surged to 80 percent. It is estimated that 90 percent of seabirds are inadvertently consuming plastic now, and that by 2050 essentially all seabirds will be.

Injury and death can result from the physical presence of plastic in the body—filling the belly, blocking passage of food to the intestine, or puncturing organs in the case of plastics with sharp edges. It can cause decreased appetite, preventing sufficient nutrition and increased levels of toxins in the body. Plastics absorb toxins from the environment, such as polychlorinated biphenyls (PCBs), and also release chemical additives from their structure (for example, BPAs, phthalates, brominated flame retardants). These toxins accumulate in tissues, resulting in reduced survival and reproduction. Before I was aware of the complexity of marine plastic pollution, my only real concern was the issue of entanglement. There is a myriad of ways that animals can become entangled in plastic along coastlines and in oceans (and in rivers and lakes for that matter), with bags and ropes and ribbon and beverage container holders—a Dr. Seuss list of human castoffs. The ear loops of disposable face masks. For turtles, seabirds, seals and whales, entanglement in plastic strands of any sort can impede movement, cause drag and decrease their mobility, resulting in starvation and death. For turtles, plastic debris on beaches is another source of danger for nesting females and emerging hatchlings.

Seabirds are also bringing plastic to shore and using it in their nests, like at Cape St. Mary's and Baccalieu Island, where gannet nests are strewn with red and green fishing rope. The island of Runde in Norway has a large northern gannet colony in an area that is under intense fishing pressure. The nests in that colony are largely constructed of both fishing lines and sections of net. Scattered through the colony, dead gannets hang, hopelessly entangled in the mesh they carried back to their nests, meant to be a safe place to raise their young. Plastic ropes, bits of gillnets and other colourful, unidentified strands are used to augment their nests of grasses and seaweed. Young chicks are curious. They will pick up and poke at the objects around them, acquainting themselves with their surroundings. Feathers. Muddy sticks. Plastic. They will eat small pieces, and there is the ever-present risk of choking, entanglement or strangulation.

But fishing gear is a special case.

Gear is, after all, designed for entrapment. In an ideal world it would capture only the targeted species for which it is deployed, and at the allowable level. Enter the real world, where mesh size, rope and hooks do not distinguish between what is desired and what is taken as collateral damage. These non-target species are referred to as "bycatch" and include other fish species, turtles, seals, whales and seabirds that pursue the fish that the net is set to catch. Populations of all albatross species in the Southern Ocean have drastically declined, for example, largely from drowning on the hooks of longline fishing gear, or in trawl nets.

Entanglement can also happen just by being in the wrong place at the wrong time—a whale migrating past a fleet of crab gear in the spring, or a puffin feeding in the same current that draws a lost net in its path. Gear routinely gets lost or is discarded, but it continues to "ghost fish," with no harvest. Just useless, wasteful death.

Fishing ropes, nets and lines were once made of natural fibres such as cotton and hemp, and ghost nets would eventually break down. But in the early sixties polypropylene and nylon came on stream, and there has been no looking back. It is cheap, durable— and strong enough to withstand the opposing forces of whale and gear without breaking. Now fabricated from polypropylene, lost gear continues to kill for years after it has gone missing, until it eventually breaks apart. Commercial fisheries are one of the main sources of marine plastic pollution in the world's oceans: 46 to 70 percent of marine macroplastic is from ghost fishing gear that has been either lost or discarded. A report by the United Nations Environmental Programme (UNEP) in 2018 indicates that over six hundred marine species are being harmed by plastic through entanglement and ingestion. Of these, 15 percent are in danger of extinction.

————

The first time I left the coastline for the open ocean was the spring I graduated from university. I was hired by Jon Lien, professor of animal behaviour and head of the Whale Research Group at Memorial University. He was world-renowned for his pioneering work in releasing entrapped whales from fishing gear. I was to join the DFO research ship *Gadus Atlantica* and conduct whale surveys on the Grand Banks for three weeks in May of 1987.

Jon was tall and broad-shouldered, with wild hair and a bushy beard and moustache. He was an American of Norwegian descent and he looked the way you would imagine a Viking to look—an affable Leif Ericsson. He was expansive; he exuded energy and enthusiasm and searched tirelessly for solutions. Funding was lean and he knew how to make it stretch, to leverage. He robbed Peter to pay Paul, as the old saying goes. He scrunched up his nose to adjust his glasses rather than using his hands. Instead, he used his hands to talk; to tie lines, cut ropes; to help you see what he envisioned. Or they were submerged, elbow-deep in something dead and rotten.

He was the kind of person who believed anything was possible if you just applied yourself, and he pulled everyone around him into the vortex of his passion. Jon saw that there was a big problem with whales getting entrapped in fishing gear. Both fishermen and whales attracted to the same places for the same reason: fish. So he started the Whale Release program.

He visited Newfoundland from the US for the first time in the late sixties, after applying for an animal behaviour position at Memorial University. The place was wild and rich; beautiful and treacherous; an ocean that offered up so much life and would just as readily take it away. It was a force to be reckoned with, and it suited him to a T. He was offered the position and he took it.

Jon and Judy Lien had a small family farm in Portugal Cove, where they grew vegetables and Judy kept goats. Judy was the yin to Jon's yang; she was soft-spoken, gentle and kind. She played classical music in the barn for the animals. The goats came to her when she called and followed her to and from the barn and the pasture. Perhaps of greatest importance to the success of their lifelong union: Judy had infinite patience.

I was hired with my friend Andrea to lead half a dozen other undergraduate students in developing portable labs for thirty schools to support a children's education curriculum about whales and their environment called Wet and Fat. We spent the summer combing beaches, collecting samples. There was one particular cove that Jon knew of in Trinity Bay that had loads of pilot whale skulls. Known locally as potheads, they are small whales that live in groups called pods, and in summer they come inshore in pursuit of squid. Before the whales became scarce in the 1960s, it was customary for locals to herd pods of pothead whales onto the beach by clanging pots together in a flotilla of small boats. Once stranded, the whales were processed on the spot; the meat and fat harvested, the bones left behind on the beach and in tidal waters. We were tasked with collecting fifty skulls and vertebrae for the lab kits. Jon supplied a van, a trailer, a four-metre aluminum boat, a small outboard motor and directions to a dock directly across the bay from the whale boneyard.

We found the small cove, a still and silent graveyard that must have been raucous, violent and bloody forty years earlier. The bones stripped clean of any remaining flesh by the diligent grazing of thousands of tiny plankton, bacteria and marine worms. Some bones bleached white by the sun, others green with algae under water, sodden and heavy. We loaded the skulls into the four-metre open boat, along with a large pile of vertebrae. The boat was unstable, the gunnels close to submerging under the weight. We made several slow, unsteady trips and eventually managed to return safely and load the bones into the van.

We didn't see much of Jon that summer. There was still an active inshore cod fishery in Newfoundland in the early eighties and his Whale Release emergency line kept him busy all day, every day. From June to August, whales and fishermen were both actively seeking fish inshore, and as far as I could tell, Jon never slept. He had his Zodiac and trailer hitched to his Mazda pickup at all times, ready to

respond to the next call for help. Jon was as dedicated to the fishermen as he was to the whales. At that time a cod trap alone cost about
$14,000; the destruction of the trap meant not only the financial cost
of the net but the loss of an entire fishing season. To lose a trap was
financially devastating. Jon understood this. And he cared about it.

Jon usually had a grad student or an assistant in tow, but he also
engaged the help of the fishermen whose nets were in peril. He
understood that it was important to involve them in the recovery of
the whale and their gear. He understood empowerment before
empowerment was a thing.

He would start by meeting the fishermen, getting the story. Not
everyone loved whales. They could rob you of your livelihood and at
close range could be, quite frankly, terrifying. Although each situation was urgent, Jon approached everyone with a sense of calm and
respect, putting people at ease. The net owners were offered the
chance to join him in the Zodiac, or they could follow his rescue
mission in a small boat. Jon was the first to attempt whale releases,
and he learned from trial and error. Before attempting to manipulate
the gear or the whale in any way, he donned a mask and snorkel,
slowly approached the whale's head and submerged his own. He
made eye contact, firmly believing that there was an essential understanding imparted between the stressed animal and the person who
was going to bring relief.

Humpback whales were by far the most commonly entrapped.
They were gentle and readily co-operated. Jon compared them to
cows—big, lumbering, docile beasts. With the assistance of his
assembled crew, he would tie a rope to one of the sensory tubercles
on the whale's rostrum—the sensitive bumps on their head. And
like the ring through a cow's nose, he would pull and push, this

way and that, while using poles and gaffs to pull the ropes off the whale. He had a knife that he used sparingly and only when absolutely necessary, to cut the net. The less damage to the net, the less downtime for the fishermen.

Fifty tons of whale that could kill with a single strike of its tail: it was nerve-racking for the uninitiated. Sometimes the release could take hours; that's a long time to hold your breath. But then the final moments—the last slip of the line, the relief, and the sheer exuberance of freedom expressed by the exhausted whale—echoed in everyone present.

Unfortunately, not all entrapments ended this way. Sometimes the demands were too great and Jon couldn't get to the whale in time. In these cases they would drown, too exhausted or too entangled to drag themselves and the gear to the surface to breathe. We had the solemn task of collecting samples from one of the whales that had succumbed like this in Trinity Bay.

We arrived at the home of the fisherman whose gear had entangled the whale. His family spoke of the haunting moans they heard all night before the whale finally expired. It had been awful to listen to, followed by a silence that was even worse. The death had occurred a few days prior; by the time we got there, the whale was hauled up near an isolated beach. The weather had been hot for several days and the whale had ballooned with gas, its decomposition accelerated by the heat. The pleats on its massive throat that ran halfway down the length of the body—meant to accommodate the over fifteen tons of fish and water taken in a single mouthful—were now expanded to full capacity. The humpback's size, once majestic, was now grotesque. We were a little shocked at the sight and more than a little shocked by the smell. We stood staring, in what amounted to a horrified stupor. Jon allowed us a moment to absorb the tragedy of it. But he also knew we had a job to do—a change in tone was necessary. Jon climbed on top of the bloated belly and jumped, bouncing as if it were a trampoline. "Anyone want to try?"

We began by collecting barnacles from the whale's rostrum, the snout an easy place to start. Whale barnacles are small crustaceans with a hard calcium carbonate shell. They attach themselves permanently to whale skin, getting a free ride while filter-feeding through the water column of the whale's choosing. Barnacles are thought to be largely harmless, although the relationship between whale and barnacle is not fully understood. Lots of possibilities fuel debate. They could be completely harmless, essentially imperceptible hitchhikers. They could confer a benefit—the hard shells with razor-sharp edges possible weaponry used in the whale's defence. Or perhaps they are an annoyance that causes the whale to breach in an effort to dislodge the intruders. Sometimes the barnacle load can become great, possibly causing drag in the water. Whatever the relationship, it is worth exploring with children and a great agent for learning more about whales, their environment and their behaviour. With sharp filleting knives, we carved through the thick skin, taking samples of both barnacle and host, and threw them into a large plastic bucket of salt water, warming in the sun.

Whales are divided into two groups, the toothed whales and the baleen whales. They have different modes of feeding and tend to spend their time at different ocean depths, and are therefore each more susceptible to different kinds of threats. The toothed whales are the more diverse group, with seventy-two species ranging in size from the smallest porpoise (45 kilograms) to the sperm whale (57,000 kilograms). Most of us are familiar with porpoises and dolphins, but there are many other species that inhabit deep waters and tend to spend less time at the surface, appearing only for a few breaths before submerging again for hours, undetected, into the black depths. Some of these species are so elusive, their existence is known only from corpses that have washed ashore, like the Cuvier's beaked whale. They pursue soft-bodied prey that are also found in deep water, primarily squid. Their dark skin scarred with perfect round circles from the suckers on the tentacles of giant squid; stab wounds from the beak of the squid's mouth—a testament to the violent battle between the two giants. The squid attaching itself to the whale's head, the whale pummelling the squid against the ocean floor with its massive bulbous head. The battle hard-won, the whale sucks the slippery victim down its throat, the lifeless tentacles giving way. Some of these battles are played out at depths as great as three kilometres under the sea—three times deeper than any navy submarine can reach before being crushed by the water pressure. Moby Dick was a sperm whale.

Baleen whales tend to be much larger than toothed whales. They feed on massive volumes of small fish or krill and have evolved an incredibly effective structure for capturing them, called baleen. Baleen grows in triangular plates that are rigid and made of keratin, the same substance as human fingernails. The inner edge of each plate is frayed, like the ragged edge of a torn brin bag. The baleen plates grow in a compact row along the upper jaw, the frayed edges acting together as a sieve—the huge mouthfuls of water pressed out from the mouth by the tongue, the small prey retained by the enmeshed fringes. Three of the large whales that are commonly seen along the Newfoundland coast in summer—minke, humpback and fin—all have baleen and feed on capelin, following them inshore.

Any of these whales could become entangled in fishing gear, and they do. But, maybe due to sheer numbers, humpbacks seem to be the most vulnerable.

The baleen had to be hacked from the jaw using an axe. Each thwack of the axe cut into the fleshy part of the whale's jaw, releasing a bubbling froth of gas and blood and an unimaginable stench of rotting flesh. The grisly attack, coupled with the overwhelming smell of rot, was unbearable and we lost a few members from our team. One was rocking herself like a baby, arms around her stomach, consoling herself from the assault on both the whale and her senses.

It was a hot, gruelling day and by the end of it we were all sweaty and rank and relieved to be done. I waded out to my knees, rinsing the blood and filth from my arms in the frigid water. I turned back toward the beach. Jon was ankle-deep and mid-thought, hands on hips. I saw my advantage, and whipped my hands through the water, soaking him with the cold deluge of splashes. Laughing, delighted by my well-calculated mischief and taking advantage of the upper hand. Jon turned—an eyebrow raised that said, *Really? You think this is how it ends?* He proceeded to walk directly toward me (*uh-oh*), scooped me up and hurled me, Viking-style, into the deeper water.

That summer there was an endangered North Atlantic right whale entangled in a cod trap in the bay at Peter's River, near St. Vincent's. This was a rarity, not only because the species was endangered but also because they normally spent their summers feeding on different prey farther south. It was far outside its normal range and had got itself into serious trouble.

The right whale is a different beast than the humpback altogether. Both whales are about fifteen metres long, but humpbacks are more streamlined, built for pursuing fish with relative speed and agility, and weigh about fifty tons. Right whales move slowly through the water, filtering tiny plankton through their baleen as they go. Their bodies are more barrel-shaped and weigh a full twenty tons more than a humpback. And if entrapped humpbacks behave like docile cows, this whale was more like a raging bull. Jon was called to the scene and Andrea and I went with him.

Cod traps are basically large box traps, anchored to the bottom, with a leader line attached to the shore. The leader line had been cut before our arrival and the whale was badly tangled, writhing in a violent panic to free itself. Held in place by the net and the weight of the anchors, the tail thrashing, it was struggling to lift its blowholes out of the water. To breathe. It was completely unapproachable in the Zodiac. So Jon called the Coast Guard to come with a larger vessel equipped with a winch, to lift the whale's tail out of the water and attempt to disentangle it that way.

It took hours for the vessel to arrive. We waited, walking the length of the expansive beach to kill time and avoid watching the whale, counting the breaths. By early evening, a vessel appeared from around the far headland. After a day of overheated pacing in rubber boots, we were finally going to be able to do something. We turned back, rushing to keep pace with the vessel's progress parallel to the beach, in order to arrive at the Zodiac as the vessel was approaching the cod trap. But Jon had other ideas. Or he forgot we were there. Or after a day of waiting, five more minutes was untenable. Or a bird in the hand was worth two in the bush. A fisheries officer had arrived in his pickup in the afternoon, having heard about the call to the Coast Guard for assistance, and he was standing next to Jon when the Coast Guard appeared. I could imagine Jon's split-second decision. *"Let's go!"*

The fisheries officer had no interest in joining Jon in the Zodiac. He was a salmon guy. Seventy tons of panicked whale was not what he had signed up for. The whale was not the only one panicking at this point.

Andrea and I watched from the shore as the Zodiac sped off with both men aboard. In crisis, I need to move, to be busy—to do. But all we could do at this point was watch and wait. Jon and the fisheries officer boarded the boat and they made the final approach to the whale together, the Zodiac tied to the stern. Jon had forgotten a tool in the Zodiac and asked the officer to go back and retrieve it while he planned an on-the-spot rescue strategy. The fisheries officer did not want to do any such thing. The whale was below the surface and he was not keen on getting in a rubber boat one-third the length of the whale, whose location was close but unknown. Nevertheless, he jumped into the Zodiac and scuttled to the back where the toolbox was stored, as quickly as the swell allowed, pitching from side to side, bending deeply so he could hold the rubber gunnel—a deep knee sprint. *Got it!* He turned to retrace the sprint, a little more awkwardly with the weight of the toolbox in one hand. But his luck ran out. Just as he turned, the whale surfaced, lifting the Zodiac clear out of the water. He dropped the toolbox in favour of holding on for dear life, aloft. It lasted only a moment before the whale submerged again. The fisheries officer scrambled for the toolbox and leapt back onto the sturdier vessel. If it had been an Olympic event, he would have won gold.

They attempted to gain control of the tail in order to hoist it up with the winch. This technique had not been tried before; they were improvising on the spot. It was not going well. From the shore, Andrea and I could hear the tail hitting the side of the ship—a loud thunderclap, resonating against the cliffs and filling the bay. The animal had to be exhausted. That it could still muster this strength was an awful miracle. The evening dragged on. The effort had to be suspended as darkness approached. By first light, the whale was gone. And with it, the cod trap. The anchors.

Chapter 7

Right Whale, Wrong Place: The Plight of the
North Atlantic Right Whale

Following the dramatic encounter with the entangled whale at
St. Peter's River, I did not see another right whale for three decades.
With the cod moratorium in 1992, interactions between whales and
fishing gear decreased significantly around coastal Newfoundland—
one of the most important inshore areas for large whales in the world.
But this did nothing for the North Atlantic right whale, which was
having different fishing gear problems of its own in waters farther
south—problems that haven't yet been solved. The North Atlantic
right whale has the dubious distinction of being one of the most
endangered large whales, with an estimated population of fewer than
350 remaining in the world.

In August of 2017, I conducted seabird surveys for the Canadian

Wildlife Service. The surveys were part of a larger collaborative project in the Gulf of St. Lawrence between the Department of Fisheries and Oceans and Oceana Canada, a branch of the international marine conservation organization. The primary goal was to explore the marine ecology of the Gulf using state-of-the-art underwater camera technology paired with visual observation from the ocean surface. And if we were lucky, we would see the vulnerable North Atlantic right whale.

The Gulf of St. Lawrence is a partly enclosed sea that is hugged on all sides by coastal waters of Labrador, Quebec, New Brunswick, Nova Scotia and Newfoundland. It has two channels that access the North Atlantic—the Cabot Strait (between Cape Breton and the south coast of Newfoundland), and the much smaller Strait of Belle Isle, with its tendril of the Labrador Current (between Newfoundland and the south coast of Labrador). The shallow waters of the Gulf are bisected by a much deeper trench called the Laurentian Channel, which flows from the St. Lawrence River out through the Cabot Strait. The constant input of cold ocean waters, the push and pull of tides and winds, and the fast-flowing freshwater intrusion from the St. Lawrence make for a rich and dynamic array of sea life.

The mission had both research and educational components and a strong media presence. It involved the coordination of two ships, the RV *Odyssey*, chartered by Oceana Canada, and the DFO research ship the *Martha L. Black*, which was the base for the DFO research scientists and for Alexandra Cousteau, a senior adviser to Oceana Canada and Jacques Cousteau's granddaughter. It also housed the ROPOS (Remotely Operated Platform for Ocean Science), a state-of-the-art underwater robot that could collect samples from the ocean bottom while simultaneously capturing still photographs and shooting high-definition film that could be live-streamed. It was the first time this kind of technology had been used to study the ocean floor in the Gulf of St. Lawrence.

I was on board the RV *Odyssey*, which carried the team assembled by Oceana Canada: marine scientist Dr. Boris Worm; a student ambassador, Isabelle Hurley, from Dalhousie University; a media team of two filmmakers from the National Film Board of Canada; and a photographer. They were shooting for an online marine sciences awareness project called Ocean School, capturing the collaborative research efforts from the ocean floor to the seabirds and marine mammals seen at the surface. Isabelle was to interview the various scientists in their roles aboard both ships and experience an oceanographic mission first-hand for the first time.

We left from Dartmouth, Nova Scotia, en route to our rendezvous with the *Martha L. Black* in the Gulf of St. Lawrence feeling incredibly excited about the opportunity to shoot footage of the ecologically rich and important Gulf of St. Lawrence using novel ROPOS technology. There were several areas of interest, particularly an area known as the American Bank. Would the right whales be there?

The American Bank is a stretch of the sea floor that extends south of the Gaspé Peninsula. The contours here vary, the shallow shelf giving way to a steep, rocky cliff that leads to a deeper ocean plateau. This area is washed in the nutrient-rich waters of the Gaspé Current that flows from the St. Lawrence estuary. The combination of varied substrate, depth and rich nutrients make it a very special place for marine life, and a highly valued fishing ground. The largest northern gannet colony in North America, Île Bonaventure, is a stone's throw away.

North Atlantic right whales give birth to their calves in the warm, shallow waters off the coast of Florida and Georgia—an ideal environment for giving birth, but there is no food there. Like most baleen whales, right whales fast during the winter months in the south and travel north in spring to feed in colder, prey-rich areas. Their favourite prey are a group of zooplankton called copepods, primarily one species, *Calanus finmarchicus*. This species has historically been abundant in the Gulf of Maine, the Grand Manan area of the Bay of Fundy and the Roseway Basin of the Scotian Shelf. All these areas have been recognized as critical habitat and have been managed for their protection.

But in 2015 scientists noticed a change in the right whale's summer migration. The ocean waters farther south were heating up. In response, copepods were moving farther north, to the colder waters of the St. Lawrence. And the whales followed.

A small building with three windowed walls, designed specifically for conducting marine mammal surveys, was located on the deck above the bridge on the RV *Odyssey*. It was affectionately referred to as "the dog house." From here, I conducted seabird and marine mammal surveys. Isabelle joined me sometimes, to watch and learn, to conduct marine mammal surveys of her own, and to interview me for Ocean School, film and camera crew in tow. Usually I work alone, in a corner of the bridge in relative peace and quiet, but this was hectic. My surveys were often interrupted by requests from the camera crew: "Could you turn?" "Would you look?" "Do you mind . . .?" "Could you repeat that?" ". . . one more time." I made the occasional fake survey entry for timeliness, deleted when the camera moved away. But I didn't mind—it was invigorating to be surrounded by so much enthusiasm. I learned to ignore the camera and settled into a rhythm—the inglorious task of systematically recording the expected seabirds and marine mammals along the northeast coast of Nova Scotia.

Most of the scientists and all the film crews were meant to board the *Martha L. Black* on the second day and observe first-hand the film footage and sea floor sampling collected by the ROPOS. But the winds were blowing over thirty kilometres an hour and the sea was choppy, creating marginal conditions for shuttling passengers by Zodiac between ships. The small boat rose and fell two metres along the side of the ship with every swell. Conditions that were tricky at best and dangerous for the inexperienced, particularly when laden with heavy and awkward film gear.

By late afternoon, the sea was finally calm enough to work with, and the transfer was successful. The American Bank did not disappoint. In real time, the camera revealed a rich diversity of marine life, forests of sea creatures clinging to rocks in unlikely shades and textures—deep purple, bright pink and yellow, green. Anemones, their delicate tentacles swaying in rhythm with the current like the outstretched arms of fans at a rock concert. Plump sea cucumbers, spiny urchins, coralline sponges. All filter-feeding from the current flowing around them. Hidden within, the predatory lair of the endangered wolf fish, striking out at unwary prey that pass too close. Starfish, brittle stars, crabs. Cod using stealth from below to feed on a school of capelin, northern gannets like a spray of bullets, dive-bombing from above—an absolute feeding frenzy. What we saw exceeded anyone's expectations.

The following day on the American Bank was sun-drenched and dead calm, the wind imperceptible with a half-metre swell. Perfect viewing conditions. The waters were teeming with wildlife—common, white-sided and white-beaked dolphins, so many dolphins that it was hard to keep up with data entry. Every whale closely scrutinized, to identify not only what it was—minke, humpback or fin—but equally, what it wasn't. Then, the staccato blows of five North Atlantic right whales travelling together. The heads rising just long enough to catch a breath, then slipping below the water again as the broad, porcelain-smooth, finless back rolls to the surface. The blow, an unmistakable V-shaped spray of mist. The languid lift of an enormous tail in the air, and then the slow-motion descent; small rivers of water cascading from the tail, signalling a deep dive. A chorus of cheers for each one.

I recorded the gannet feeding frenzy and then abandoned the dog house to focus on the right whales. I was as absorbed in the thrill as much as anyone, but after a while I stood quietly back from the group and reflected on the immense significance of what we were seeing, and to bear witness to something that may soon disappear forever. This novel feeding area was putting right whales in the direct path of threats they had not encountered before: heavy marine traffic from large ships and a high concentration of fishing gear. We had eighteen right whale sightings that day—a staggering number for such a rare species. By the end of the 2017 summer season, seventeen deaths from fishing gear entanglements and ship strikes had been reported.

Encounters with crab gear pose the greatest threat to right whales in the Gulf of St. Lawrence. Crab are caught in pots that are placed on the ocean floor and set individually, or strung together in a long line, or fleet. The pots are joined together by ground lines and marked by floating buoys, indicating the beginning and end of the fleet. The buoys are secured to the fleet by a line that floats vertically in the water, anchored to the bottom and attached to the first in the series of pots. These vertical buoy lines that extend from the surface to the ocean floor are the biggest problem. As the whale swims, it hooks its pectoral fin, tail stock or mouth into the floating vertical line and starts to drag it. In an effort to free itself, it may roll and dive, becoming even more hopelessly entangled, wrapping itself in the lines and dragging the weights and entire fleet with it.

Similarly, right whales are threatened by lobster gear, particularly along the migration route in the US. This sort of entanglement seems unlikely, but in areas where whale activity intersects with intensive fishing activity—like the New England lobster fishery and the snow crab fishery in the Gulf of St. Lawrence—the chances are dangerously high. There are over a million lobster pots and their associated lines active in right whale habitat, creating a situation that is completely unsustainable. Starting in 2017 there has been a sharp spike in mortalities, referred to by NOAA as Unusual Mortality Events. An estimated forty-eight right whales have died between 2017 and 2022.

Right whales are visually distinguishable from each other by irregular patterns of thickened skin on their rostrum, called callosities. This skin becomes inhabited by small white crustaceans, creating a unique pattern. Some whales are known affectionately by nicknames given by researchers who have been tracking them for years. Monitoring right whales in the North Atlantic can be a heartbreaking business. In 2020, an eleven-year-old male named Cottontail was spotted towing a heavy load of fishing gear, which was wrapped tightly around his head, cutting into his mouth and trailing behind him. Several rescue attempts failed, and after four months he was found dead off the coast of South Carolina, emaciated and severely injured. Another dead right whale was found floating far out in the open Atlantic, entangled in gear that was traced to a fisherman in Cape Breton, Nova Scotia. The fisherman was devastated by the news. There are no "bad guys" here—just a tragic clash of fishers and whales, both trying to survive. But one thing is abundantly clear: it is past time for another gear innovation.

Luckily, there are efforts afoot. Marine scientists from government and non-government organizations in both Canada and the United States are looking at technologies to innovate the crab and lobster fishing gear, focusing on the lethal vertical line. A low-tech solution involves a weak vertical line that breaks more readily than lines currently in use. A recent study from Massachusetts demonstrated that lines with a breaking force of 770 kilograms could work effectively in the near and midshore fisheries to protect right whales. Although this is very good news for right whales, there are some immediate problems that come to mind with respect to ocean plastics.

Marker buoys are designed to float on the surface. They were once made of blown-glass balls but are now made of plastic. When the vertical line breaks, the buoy and the remaining rope it is attached to are set adrift on the ocean surface, joining all the other floating ocean detritus. The rest of the gear is left behind, unmarked and difficult if not impossible to relocate. The pots continue to capture lobster or crab that will never be harvested, until the gear structure eventually fails. The remaining broken vertical line, the pots and the line that held the fleet together, slowly breaking up.

There is another option currently under investigation, that is much more promising for both right whale and ocean health. In Canada and the United States, scientists are looking at ropeless technologies using acoustic release systems and GPS rather than surface-floating buoys to detect and recover gear. Under these systems, the buoy is submerged and attached to the ocean floor. The set traps are relocated using GPS. Once near the net, a radio signal is sent at the surface, activating the receiver to release the buoy, which then floats to the surface.

In response to the high rate of right whale mortality in the Gulf of St. Lawrence since 2017, the Canadian government introduced fisheries restrictions and imposed a speed limit of ten knots (18.5 kilometres per hour) in specified zones to protect right whales. These measures appeared to be effective; there were no right whale deaths in Canada in 2018. The restrictions were tweaked in 2019 and ten right whales died, nine in Canadian waters and one found in American waters but entangled with Canadian fishing gear.

The mortality rate continues to climb. In July 2020 the International Union for Conservation of Nature, the global authority that ranks species' risk of extinction, increased the right whales' risk from Endangered to Critically Endangered. With a population of fewer than 350, the window of opportunity for recovery is closing.

There are some promising developments. The American Bank in the Gulf of St. Lawrence had been considered for designation as a Marine Protected Area (MPA) since 2011. Marine Protected Areas are federally designated areas of ocean that are legally protected and managed for long-term conservation. Activities within MPAs can be restricted or prohibited if necessary. It was finally given MPA status in 2019, supported by the richness and diversity revealed by ROPOS and reflected by the activity at the surface—seabirds and marine mammals. This can only be good news for the right whale.

In 2016, changes were made in the American Marine Mammal Protection Act that set limits on the amount of bycatch associated with fish imported from other countries, including Canada. This includes the entanglement of right whales in crab and lobster gear.

A promising step in gear innovation occurred in the summer of 2022, when a new ropeless gear technology underwent trials on crab and lobster traps deployed in Newfoundland waters. The effort was a joint venture between Jasco Applied Science (the company that developed the system), the Washington-based non-profit Sea Mammal Education Learning Technology Society and fishers from the Miawpukek First Nation. The traps were equipped with an acoustic release system and a series of floats that could be triggered to bring the gear to the surface—the same type of technology used for oceanographic moorings, on a smaller scale. This system eliminates the vertical line to the surface found in traditional gear.

As would be expected, there has been mixed reaction among fishers to the proposed and enforced changes to both fisheries closures and gear innovation. Change is always viewed with some level of fear and skepticism. On the other hand, no one wants to lose gear, fishing opportunities and markets. And no one wants to wipe out a species. We are on the razor's edge of doing exactly that.

This dire situation is not unprecedented. The vaquita, a porpoise that is the smallest of all the cetaceans, is on the brink of extinction. It grows to a maximum of five feet and lives in murky waters of shallow lagoons at the very northern tip of the Sea of Cortés, Mexico. In the 1930s the population was estimated at five thousand, but numbers plummeted due to the illegal gillnet fishery for the highly prized totoaba, a large fish sold to the Chinese market. The fish's swim bladder is used in Chinese medicine and can command a whopping $46,000 per kilogram on the Chinese black market. For this reason, it is known as "the cocaine of the sea." The mesh size of the illegal gear used to fish for totoaba is perfectly suited for incidentally capturing vaquita. Research careers have been devoted to trying to save the species from extinction. Yet in 2018 the International Committee for the Recovery of the Vaquita (CIRVA) reported a population size of only 6 to 22 individuals remaining. If there are closer to 20 still alive, recovery is considered possible by the most optimistic, but only if the illegal gillnet fishery is eliminated immediately. Some whale scientists think it is already too late. In 2022, surveys could confirm only 10 of the tiny whales, and the illegal fishing gear remains in the water.

The North Atlantic right whale may have even less wiggle room for population recovery than the vaquita. Both species are slow to recover, but the right whale much more so. The vaquita has its first calf at about five years of age and can have a calf every other year during its lifespan—about twenty-five years. The right whale doesn't reach sexual maturity until about ten years of age (Cottontail was just getting there). They are capable of giving birth every three to four years and living to around seventy years, but that is not what is happening. Females are only living on average to forty-five years and giving birth only every six to ten years. Males are surviving longer. Scientists believe that the additional stress of entanglements is the reason females are giving birth less often and dying younger. If the right whale population declined to the current population of the vaquita, extinction would be inevitable.

I am not trying to imply anything sinister about the crab and lobster fisheries of the Atlantic coast and the Gulf of St. Lawrence. But fishers are reluctant to change and politicians like to get re-elected. All the research in the world is pointless unless it results in the changes that are required and *when* they are required. For the vaquita—and the North Atlantic right whale—we can only hope that time has not already expired.

Recent events do not offer much encouragement for right whales if changes aren't implemented quickly. According to NOAA, since 2017 the estimated population has dropped from fewer than 400 individuals to fewer than 350 in 2022. During this period, right whales have been dying faster than they have been reproducing. "Snowcone" was spotted in March 2021 dragging hundreds of feet of thick, heavy rope cutting into her mouth. She was so compromised that she was expected to die. Through several delicate rescue efforts, more than three hundred feet of rope were cut from the train she was pulling. It turned out that, through all this, Snowcone was pregnant. She somehow managed to give birth in this condition and was spotted with a newborn calf on December 2, 2021. On January 13, 2022, wildlife officials in Florida were able to closely document the entanglement; fishing rope was coming out of both sides of her mouth and embedded in her jaw. On July 26, 2022, Snowcone was sited somewhere off New Brunswick, the lines still streaming from her mouth. The calf—which should still be nursing—was gone. This is the story of one of only two known females that gave birth in 2021.

After four days of surveys, interviews, filming and spectacular underwater footage in the Gulf of St. Lawrence, the mission ended. Everyone was in high spirits—the mission had been an unmitigated success. The RV *Odyssey* continued on to Sydney Harbour and dropped off the film crews and all the collaborating teams. The air of celebration wafted gently away on the light southwesterly breeze. I remained on board in the quiet for one final day of surveys on the return trip to Dartmouth.

After all the activity and constant companionship, the solo journey along the coast offered a time for peaceful reflection. To absorb what I had just seen and experienced. The Gulf of St. Lawrence, an almost entirely enclosed sea unto itself, fed by the confluence of the fresh waters from the St. Lawrence River, the cold ocean waters from the Labrador Strait and the warmer waters of the Cabot Strait. The swirling and mixing of these waters, generated by so many interacting forces—temperature, salinity, wind and tide. To have witnessed first-hand the world created by these currents and the undersea life they support. Why both fishermen and right whales are drawn to the same location and how catastrophic that interaction has been. And to have seen North Atlantic right whales swimming, diving, feeding—. And at least in that brief moment in time, thriving.

PART

Three

Ocean Currents and
the Plastic Crisis

Chapter 8

Kittiwakes and the Japan Current

In the summer of 1988, I went to work on a seabird colony on Middleton Island, nestled in the armpit of the Gulf of Alaska seventy-five kilometres off the coast. It was my first time working on a colony outside of Newfoundland and under the influence of oceanic conditions outside of the Labrador Current. I knew nothing about the place except that, as far as latitude goes, it was slightly farther north than Nachvak Fiord. But it was a seabird colony in Alaska—the land of deep-azure skies shot through with snow-covered mountain peaks. Some version of Nachvak, except with seabirds, I figured. That was enough for me.

John Piatt, the graduate student who had introduced me to beached bird surveys several years earlier at Cape St. Mary's, was now a research scientist for the US Fish and Wildlife Service in Alaska. Based on his recommendation I was hired by Scott Hatch, lead scientist on the Middleton Island studies.

I needed several critical details in order to prepare. So I asked Scott a few questions: What species would we be studying? Black-legged kittiwakes. Their breeding efforts had been failing in Alaska for years, and this was part of a long-term breeding behaviour study to shed some light on what was going wrong. Is there any power or running water? No. Are there any trees? One or two. One or two? This bewildered me. What I really wanted to know was whether we would be situated north of the treeline, which would suggest Arctic conditions—and indicate the type of climate to prepare for. I expected it to be, given its latitude and believing that latitude was in the driver's seat as far as climate goes. But that's not what I asked.

Scott Hatch was as good as his word. As our small aircraft broke through the low grey cloud cover and circled the island, I took in the landscape. Counted the trees: two. Saw a rusty, hulking shipwreck permanently listing on the beach (what? how?). There was a functional, modern-looking building at one end of the island, a cluster of derelict houses at the other and a short gravel runway between them. An old radar tower and some other dilapidated outbuildings spoke to the brief period when Middleton Island served as an air force station. Other than these features, it was a pretty flat, nondescript island covered in grass and encircled with sandy tidal flats. And although I knew nothing about the place, I did know that kittiwakes breed on sheer cliff faces. There were no dramatic cliffs here—it was a glorified sandbar. As far as first impressions go, Middleton Island was a bust.

At first blush, it was hard to imagine where a kittiwake colony could be. And no wonder—they were nesting in some very unlikely places. On grassy slopes, completely unprotected from predators. Glaucous-winged gulls responded in kind, strolling up the slope and picking off eggs and chicks at their leisure, the kittiwakes no match for the large gulls. They were having better luck on the rusting hulk on the beach, which had once been a World War II anti-aircraft supply ship. Much of our work, Scott told us, would take place inside that forbidding, corroded wreck.

In 1942, the SS *Coldbrook* had been pursued by what was assumed to be a Japanese submarine. It may or may not have been shot before running aground on the shoals of Middleton Island. The dramatic force of the 1964 earthquake subsequently raised the shoals out of the water, creating a new beach. The ship sat on top of it, its hull lapped by the tides—and a kittiwake nest in each porthole. An unsettling testament to how badly things can go wrong. And to how adaptable and resilient nature can be in the face of unusual circumstances.

We set up shop in one of the abandoned houses. A permanent crew of four, we each had our own bedroom and shared a kitchen, replete with Coleman stove. It would be weeks before I stopped automatically flicking the switches in response to poor lighting. I ignored the bathroom, with its bathtub, toilet and sink—a reminder of the conveniences that would have been afforded if any of it worked. In stark contrast, our toilet consisted of a hole in the ground, a roll of toilet paper, a basin of water and a bar of soap. Rain gear, as needed.

We were attacking the kittiwake breeding problem on two fronts. The first involved classic field observations from a blind set up near the nesting slope, where we recorded breeding behaviours and chick-rearing success. The second part used the ship as a sort of on-site laboratory. The stern of the ship had been ripped off by a storm, leaving the engine room exposed, and kittiwakes were nesting all over the exposed pipes. The idea was to set up scales under these nests that were connected to a solar-powered computer that registered all changes in nest weight. We would take turns hiding behind one-way glass from a cramped room in the bowels of the ship and record behavioural activity that would help explain these changes—parents landing and delivering food, chicks pooping.

We entered the *Coldbrook* through a gaping hole on the port side and climbed a ladder from the hull to the first deck, nine metres above. The inside of the ship felt like the set of a horror movie: the oppressive darkness flecked with pinpricks of light through the lacy rust holes; the cold and dampness from decades of rainwater and moist salt air seeping to the marrow. Hard hats were our protection against the tons of concrete that were collapsing on disintegrating metal support beams. Using headlamps, we walked slowly and carefully along crumbling floors that could—and did—give out underfoot. A cold draft came from everywhere and nowhere—rising from the torn hull, the holes in the rusted floor, the walls. Winding down halls from open portholes. Surrounding your body like a cold whisper, lifting the fine hairs on your legs, your arms, the back of your neck.

Silence echoed in the hollow centre of the ship, occasionally pierced by the sudden scream of a kittiwake, a jolt of electricity to the heart. Near our observation room, we had to pass a meat locker, where the mummified corpse of a cow still hung, motionless. If you looked the wrong way, it would flash past your singular beam of light, the image burned on your retina. Shipwrecks, I understood, involved much death and drowning. Where were the other bodies? Every time I entered the ship, I was leaden with dread. Carefully and slowly we picked our way along hallways half-submersed in stagnant rainwater and lined on each side with cabins.

What was intensely unpleasant for us, though, was the very definition of puffin luxury. Tufted puffins were now the occupants of the crew's accommodations. Normally, puffins nest in deep burrows—long chambers in the cold earth that they have to excavate themselves. But here the work had been done for them. Each dresser drawer was occupied by a puffin pair, a small nest built inside. The hallways saw regular traffic, as birds entered and left their burrows in pursuit of food and daylight, paddling, carefree, to and fro.

The island turned out to be full of strange surprises and hidden treasures. I spent a lot of time combing the beaches; the sand spits and tidal flats that appeared relatively barren from the air were alive with animals that were completely exotic to us. Steller's sea lions loafing on the beach and in the water, swimming parallel to us, spying as we walked. Northern phalaropes spinning in circles, stirring up tiny crustaceans to eat from the bottom of tidal pools. The non-stop chatter of oystercatchers, their pale-pink legs clashing with bright-red bills. Flocks of migrating shorebirds and waterfowl. As early spring progressed into summer, activity blossomed.

One of the most intriguing things about the beach was the flotsam that washed up on its shores. Impossibly delicate blown glass balls used for floating nets, hard hats, an array of random debris inscribed with Asian characters. Nothing like the shotgun shells and beef buckets I was accustomed to seeing in Newfoundland. I couldn't fathom how these objects had found their way to a remote island in the Gulf of Alaska. At twenty-four, I was content to leave it a mystery.

Our field season ended a few weeks early that year. The kittiwake breeding season had failed once again and there were no chicks in our study plots to monitor. Only a few chicks on the whole island had survived long enough to fledge from the nest, the telltale black M on their back readily distinguishing them from the adults. It was a stark contrast to the many hundreds I was used to seeing at this time of year in Newfoundland colonies. Poor breeding success on Middleton Island could easily be blamed on glaucous-winged gulls. But what about the other struggling Alaskan colonies? Were the ocean conditions changing in the North Pacific?

The puzzling presence of those glass globes and hard hats on the beach stuck with me. It was decades before I even heard of the Japan Current. Or the Great Pacific Garbage Patch.

———

Ocean currents are central to life on earth. They explain the location of all the richest fishing grounds in the world; why fires have been raging in California; how a balloon from a six-year-old's birthday party in Manhattan can end up in the stomach of a whale off the coast of Norway; or how a disposable diaper finds its way into the diet of a polar bear on an island at the northern tip of Labrador. They explain the distribution of plastic in the ocean and are intricately linked to climate change as they distribute the heat absorbed by the ocean from the atmosphere. They explain the circulation of water around the planet, driven by winds, tides, water temperature and density, and the very origin of over half the air we breathe. We call the ocean by different names, depending on the continents that rise above its surface, interrupting its flow. But in essence, there is really only one global ocean, and all the water in it gets circulated around the earth by a few major ocean current systems. The Labrador Current—"my" current, whose purpose and effect I understand at a distinct, regional scale—is an integral part of a global current circulation. And the engine that drives the system originates at the poles.

To understand how the engine works, you need to know something about the Arctic and Antarctic cold bottom water, which drives the circulation of the deep ocean currents around the entire planet. The cold, deep currents are aptly described as the ocean's conveyer belt. The engine that runs the conveyer belt is the formation of the very dense, cold waters of the polar regions, born from the formation of sea ice.

As sea ice starts to freeze, ice crystals form. These crystals begin to consolidate and eventually create a continuous sheet. Salt doesn't freeze, so as the ice crystals form, the salt gets pushed out, forming channels in the sheet, and creating a dense, salty brine layer beneath it.

Sea water freezes at a temperature of about minus 1.8°C—lower than the freezing point of fresh water, because of its increased density from the salt content. The greater the salt content, the lower the freezing point. The briny layer created from sea ice formation is extremely salty and remains a supercooled liquid.

Another important feature of cold ocean water is that it is very oxygen-rich. This is because oxygen dissolves more readily in cold water; the colder the water, the greater its ability to retain oxygen.

The resulting very cold, salty, oxygenated and dense water that has been created from sea ice formation sinks to the ocean bottom and, under the right conditions, becomes the Arctic and Antarctic deep bottom waters—the source of the deep ocean currents, or the conveyor belt. As these frigid waters move away from the poles and interact with warmer ocean bodies, they lose strength, but become recharged as they reach the opposite pole and start the return trip. The deep-water currents are very slow; it can take a water droplet 1,000 years to complete the circuit.

As geographically complex as the land above, the ocean floor is characterized by deep canyons, mountains, ridges and valleys. The course of the deep ocean currents is dictated by the contours of the ocean floor, flowing off the shallow polar continental shelves—the underwater extension of the continent itself—and sinking thousands of meters below. The force of the current is driven by thermohaline circulation—differences in water density that are controlled by water temperature (*thermo-*) and salinity (*haline*). These currents become, essentially, deep rivers that interact with ocean bottom features and other, warmer ocean water bodies, to profound effect.

When a deep current encounters an obstacle—like a sea mount or a continental shelf—the impact forces a wall of cold water toward the surface, bringing with it nutrients from the bottom and lots of oxygen. This is where magic happens. The nutrient-rich upwelling water is exposed to light, which creates conditions for phytoplankton to thrive.

There are hundreds of species of these single-celled algae, some with the delicate and intricate structure of a snowflake. They are small, but they are mighty. Under the right conditions they can form blooms that extend over hundreds of kilometres. Through photosynthesis, they use light energy and dissolved carbon dioxide from the atmosphere, releasing oxygen back into the air. The carbon becomes part of the phytoplankton's cell structure, creating carbohydrates that animals can use for food. Then one of two things happens: either the phytoplankton die and sink to the ocean bottom, taking the carbon with them, or they get grazed on by zooplankton. In the latter case, the carbon that is passed on is then incorporated into the zooplankton's own bodies, or is passed through it, the waste also destined primarily for the ocean floor. The carbon continues to get passed along by other marine animals at various ocean depths that in turn create waste and die, releasing carbon in the process. If you have ever seen a brown, fetid cloud trailing behind a well-fed whale, you will know that large volumes of carbon and other unmentionables are being returned to the surface and cycling through the system again, joining the other marine detritus, known as marine snow, drifting toward the ocean floor. This cycling of carbon is referred to as the ocean's biological carbon pump; it transfers a whopping 10 gigatonnes (10 billion metric tons) of carbon from the atmosphere to the deep ocean every year and plays a critical role in mediating climate change.

Over billions of years, it is the accumulation and decomposition of these carbon-rich particles that forms oil deposits—fossil fuels. When fossil fuels are brought to the surface and burned, the massive stores of carbon that have been removed from the atmosphere over the billions of years are released back into the atmosphere.

Phytoplankton are not only integral to the ocean's biological carbon pump, they are responsible for over half the photosynthesis on earth. And over half the oxygen we breathe.

While particles eventually end up on the ocean floor, the journey can take hundreds of years; it is usually not a direct path from surface to sea bottom. There is an important series of currents that swirl at the ocean's surface as a result of the earth spinning on its axis, and these currents affect the entire planet. In response to the earth's rotation, trade winds flow from east to west along the equator, pulling water at the ocean surface with them. The globe's rotation causes the winds to spin off in an arc, to the right in the northern hemisphere and to the left in the southern hemisphere—a phenomenon known as the Coriolis effect. The ocean's surface waters at the equator follow suit, flowing westward with the wind then ricocheting off the continental shelf, creating spinning ocean gyres to the north and south. There are five of these enormous gyres in the world's oceans, one for each ocean basin: the North and South Pacific, the North and South Atlantic, and the Indian Ocean. The five gyres all behave in exactly the same way: as they spin, they concentrate everything that is free-floating toward their centre, like the tea leaves that gather in the centre of a stirred cup. The impacts of these gyres are far-reaching and more significant than you might imagine.

Coastal plastic is carried out to sea, ebbing and flowing, passively responding to the demands of tides and wind. Some of it sinks along the way, but much of it is drawn into the nearest gyre, swirling and slowly gathering at its centre. The content of each gyre depends very much on the coastlines that feed it.

Take the North Pacific Gyre. It is bound along its edges by four major currents. The southern edge is bounded by the equatorial current that flows westward with the trade winds (the North Pacific Equatorial Current). When the current bounces off Asia's continental shelf near Japan, it spins northward, becoming the Japan Current, then the North Pacific Current, which heads east until it is forced south by North America's continental bulk and becomes known as the California Current. Of all the gyres, this one has by far the greatest concentration of plastic debris. This dubious distinction is a result of the fact that it conveys water from east Asia, which has some of the most populated and plastic-choked coastlines in the world. At the centre of the gyre is an area of 1.6 million square kilometres (about twice the size of Texas) with somewhere between one and 3.5 trillion plastic pieces floating within it. You may have heard of it—it's known as the Great Pacific Garbage Patch.

The Great Pacific Garbage Patch is not a vast mat of floating debris as far as the eye can see, as its descriptor might suggest. For the most part, it is invisible to the naked eye. It is made up primarily of tiny plastic particles, called microplastics, which form a toxic soup, aptly described as plastic smog.

And then there is the Sargasso Sea. Look on any map, you won't find it. I had heard of it in my childhood, snorkelling in a lake at our family's cabin. Watching from above, the creepy eels like dark shadows, almost undetectable, slithering between rocks along the bottom. I was told these horrifying and fascinating creatures were born in the mythical Sargasso Sea (does it really exist?) and mysteriously wound up at the bottom of our lake. This sounded like a plausible origin for a sinister-looking bottom dweller to me. And that was the full extent of my knowledge of the Sargasso Sea and its inhabitants for a very long time.

The Sargasso Sea is not defined by land masses but by the pull and swirl of ocean currents. It is named for the seaweed *Sargassum*, a free-floating algae that is carried with it and concentrated in floating mats that can extend for many kilometres. For hundreds of years sailors feared it, believing it capable of ensnaring ships. The Sargasso Sea now has a much less treacherous reputation. Known as "the golden floating rainforest," it is a rich source of zooplankton, a safe haven for tiny creatures, the place where American eels return to spawn and die, and a resting spot for all manner of migratory species. It forms a solid base for a marine food web, and provides for animals that are farther up the food chain, including seabirds and whales. *Sargassum* is a life raft.

I have sailed along the edge of the Sargasso Sea and seen phalaropes poking and probing in mats of seaweed, picking up small prey with their needle-thin bills. I was not in an area where the *Sargassum* was at its thickest, and I wondered at the hidden forces that guided phalaropes to such a rich food source in the middle of the open ocean. The *Sargassum* was not a random find. Phalaropes have likely been relying on it since the beginning of phalarope migration. Since the beginning of phalarope time, basically. This is the same shorebird I had seen in tidal pools on Middleton Island, Alaska, spinning in circles, creating tiny gyres of their own to bring bite-sized crustaceans to the surface. The phalarope, it would seem, has mastered the spinning gyre.

The Sargasso Sea itself is at the centre of the North Atlantic Gyre, the North Atlantic equivalent to the Pacific's Great Pacific Garbage Patch. And like the arm of the Gulf Stream that flows northward and becomes the North Atlantic Current, so too does the current in the Pacific. Its dynamic equivalent is known as the "gulf stream of the Pacific," or the Japan Current.

While the Sargasso Sea concentrates *Sargassum* and the extended food web that it supports, it also concentrates threats to that marine ecosystem, including plastic and other toxins—and the threat of bycatch from fisheries that are active there too.

I wonder not only at the miraculous pairing of this tiny seabird with this floating refuge, but also at how much microplastic the phalarope is ingesting, either by accident or by choice, mistaking small particles for its tiny prey. As more plastic enters the ocean, will the Sargasso Sea, like the North Pacific Gyre, become more refuse than refuge?

The ocean around the equator absorbs a great deal of atmospheric heat from the intense sun exposure there, making water temperatures warm in tropical regions. This makes sense, but it doesn't hold everywhere. There are areas of much colder water in tropical regions, and here's why. The water that is pulled away from the east side of the ocean (and therefore the west side of a given continent) must be replaced. The cold, nutrient-rich ocean bottom waters rise to the surface along the eastern side of the ocean to displace the water that is being moved west, creating a cold-water upwelling. As these cold waters are drawn west, they start to heat up along the surface as well. By the time they make it to the other side of the ocean, they are part of the warm water surface current that feeds the gyres that spin north or south, depending on which side of the equator the gyre is on. The intensely warm water on the west side of the ocean warms the air above it, which rises and condenses, forming low-pressure systems with rain clouds. Not so over the colder waters in the east, where high-pressure systems prevail and result in more arid conditions. This is why deserts tend to be on the western side of continents near the equator. This is also why—thanks to the nutrient-rich cold upwellings—these are the regions where marine life is rich and productive fisheries occur.

The effects of the trade winds on ocean temperature and currents are the greatest in the Pacific Ocean, the largest ocean on earth. Here, the surface water has longer to gather heat as it moves west, creating a greater temperature gradient between the cold east and warm west sides of the ocean.

The cold-ocean upwelling along the tropical region of western South America creates the Humboldt Current, one of the richest marine ecosystems in the world. This is the home of the humble anchovy. Like the capelin of the north, it is a small schooling fish with an astonishing biomass that supports millions of seabirds and whales and is responsible for one of the world's largest fisheries.

There are years when the trade winds that drive the cold-water upwelling weaken. Without the momentum of the wind, the warm water mass that normally gets pushed west moves very little. Without warm water being pushed west, there is very little displacement by cold deep-ocean upwelling waters in the east. The lack of cold-water upsurge leads to the collapse of anchovy populations—the key foraging species for millions of seabirds and other animals—which is devastating for marine life and fisheries in the regions. This is the El Niño phase of the climate phenomenon known as the El Niño Southern Oscillation (ENSO). *El Niño* means "the little boy" or "the Christ child" in Spanish, because the greatest impact of El Niño occurs around December. El Niño generally manifests every two to seven years, although the frequency seems to be increasing.

At the other end of the spectrum is the La Niña (the little girl) phase, with opposite effect. La Niña occurs when trade winds are particularly strong, which intensifies their normal effects, and the volume of water moved along the ocean surface is greatly increased—more heated water buildup in the west and more cold-water upwelling in the east. The result is more rain in the western Pacific and much less in the already drier eastern Pacific. The year 2020—which saw uncontrolled wildfires raging in southern California—was a La Niña year. It has lasted longer than usual, continuing through 2021 and is forecasted to extend through the 2022/23 winter. And with it, more droughts and a longer fire season for California and beyond. Between years of El Niño and La Niña there is a neutral phase.

There is not an equivalent to the Pacific's ENSO phenomenon in the smaller Atlantic Ocean, but the trade winds still have a very powerful effect on its currents. In fact, the five gyres of the ocean basins, all driven by the trade winds, are dynamically the same. The

Gulf Stream is formed from cool waters that are carried by the trade winds, just above the equator off the west coast of Africa. The current moves west, absorbing atmospheric heat along the way, until it reaches the Gulf of Mexico. At this point it is a very warm current and officially becomes known as the Gulf Stream. From here it veers north, bringing heat with it, displacing and mixing with other, colder water and affecting the climate of three continents. Its dynamic equivalent in the Pacific Ocean is the Japan Current.

The Japan Current is also known as the Kuroshio Current, meaning "the black stream" in Japanese. The current morphs into the northern North Pacific Current, conveying hitchhikers that travel with it along offshoots that reach well into the Gulf of Alaska. This explains the mysterious presence of a Japanese hard hat and all the other exotic detritus on the remote coastline of Middleton Island.

As for the kittiwakes, despite their poor yearly breeding success, the resilient seabirds continue to nest on the island and seem to be thriving well enough there. The decades of research led by Scott Hatch has revealed something interesting and unexpected: Pacific kittiwakes have much lower breeding success than their Atlantic counterparts, but they live twice as long. As is often the case with good science, this finding raises more questions than answers.

Chapter 9

Following Currents to the Ends of the Earth

In the spring of 2019, I accidentally signed two long back-to-back expedition cruise contracts with the Norwegian company Hurtigruten. This meant that, from March to May, I travelled the ocean from Antarctica, along the west coast of South America, through the Panama Canal and up the east coast of North America, before being dropped off at home at St. John's harbour—a journey roughly the same length as the annual one-way migration of an Arctic tern. I had already been to Antarctica in 2017—a trip I will say more about later— but this fortuitous mistake offered a new and incredible opportunity to see the shifts in both human and marine life as we travelled through cultures and currents, and to observe how they encountered and interacted with each other. I flew to the port city of Ushuaia in southern Argentina to join the ship and set out for Antarctica.

The continent of Antarctica is entirely surrounded by a single, open ocean. The Pacific, Atlantic and Indian Oceans join forces, driven by a ruthless westerly wind that encircles the continent and sweeps a power-ful ocean current with it, called the Antarctic Circumpolar Current.

The harshness of the Southern Ocean is long-known and feared. Its immense richness and remoteness has attracted whalers, sealers and explorers, keen to exploit the treacherous frontier. The force of those westerly winds became notorious, whipped up to greater strength the farther south you went. With increasing latitude they were known as the roaring forties, the furious fifties, the screaming sixties. There was a common adage among sailors: "Below 40 degrees south there is no law. Below 50 degrees south there is no God." So, yes. Harsh.

The Antarctic Circumpolar Current is the longest current on the planet and the most powerful; it moves a hundred times the volume of all the world's rivers *in one second*. The current acts as a barrier against warmer surface waters to the north, creating a closed and frigid system around Antarctica. Here, you have cold water whipped around the surface and even colder, dense water moving slowly northward at the bottom. The average thickness of the ice sheet over the continent is just over two kilometres. At its thickest, the ice cap is 4.8 kilometres thick.

Sea ice starts to form in the fall, growing five kilometres each day. At its maximum, the annual ice sheet covers about 18 million square kilometres, essentially doubling the area of Antarctica and earning its reputation as the coldest, windiest place on earth.

At its northern limit, the Southern Ocean butts up against warmer waters in an area known as the Antarctic Convergence, a dynamic zone where opposing ocean currents and temperatures collide. It is home to an incredible array of seabirds, many of which are found nowhere else in the world. And for me, it is one of the most compelling places on the planet.

We sailed through the silence and space of Antarctica, sparsely and briefly visited by people, until it gave way to the vibrant colours and cultures of the Chilean coast. These changes were accompanied by a gradual morphing of seabird composition, the Antarctic breeders exchanging with those that prefer slightly more temperate conditions. We left behind the ice-dependent Adélie penguins completely; the gentoo and macaroni penguins were willing to make a go of it in the somewhat more temperate waters across the Drake Passage to the southern tip of Argentina and Peru. Some of the ship's followers stayed with us for the whole crossing—the Cape petrels, the giant petrels, now old friends. Others handed off the baton at the Antarctic Convergence, not quite ready to give up the Antarctic summer. The snow petrels petered out, giving way to greater shearwaters and sooty shearwaters, which gathered in the thousands off Cape Horn. The light-mantled albatross became the black-browed. And with the warmer ocean that came with crossing the Antarctic Convergence, the air gradually warmed too.

For days we travelled north in the cool Humboldt Current, still connected to Antarctica by the whisper of its cold waters, drawn up the coast of Chile and Peru, replacing the rush of water blown west by the trade winds. The richness the cold water brings is evident in the sheer abundance of life—thousands of shearwaters, petrels, black-browed albatross, Magellanic penguins. The species continued to turn over as we moved north, grey-headed albatross slowly being outnumbered by their close relative the Chatham albatross. Peruvian boobies, a species that is a hallmark of the Humboldt Current, appeared. Elegant from a distance, goofy-looking up close, it is a close relative of my beloved northern gannet and has the same long wings and tapered tail. The same close-set eyes staring forward that give it a permanently stupefied look. Though the effect is comical, the eye placement is necessary for depth perception and essential for a plunge-diver. The Peruvian booby torpedoes into the water after anchovies and sardines the way gannets do for capelin and other forage fish. I had always wanted to see boobies. And boobies I would see.

The current also bathes the shores of the Galapagos Islands. It is the reason the Galapagos penguin even exists. And the Humboldt penguin speaks for itself. Penguins are birds that thrive in cold southern oceans, right? We think mostly of Antarctica when we think of penguins. A penguin in the tropics, just north of the equator, even—it doesn't seem possible. But the Humboldt Current brings a bit of Antarctica with it—enough to satisfy the needs of a Galapagos penguin.

The ship made a shortcut from Chile to Peru in the northern portion of South America where the land mass fans out, taking us into deeper waters off the coast and out of the Humboldt Current. The water warmed and the wildlife grew sparce, reduced to a few transoceanic migrants and masked boobies, diving for fish that would be few and far between if not for the herding by tuna, below. The water depth confirmed by the appearance of sperm whales.

The following day we turned toward shore again, the waters cooling as we approached, the birds gathering in huge concentrations. The year 2019 was marked by a mild El Niño, and the equatorial waters of the Pacific were a little warmer than usual. I spotted a few Humboldt penguins a much greater distance from shore than expected. Having to move farther to find fish, they were seen as far as 90 km off the coast—a good 60 km beyond normal foraging distance.

We made landfall near the Ballestas Islands, and made our way to the small port town of Paracas, which splits its efforts between the fishery and the tourism generated by boat tours around the islands. The port was blocked with fishing boats, splashes of gasoline iridescent on the surface. Careless bits of plastic trash, cigarette butts and empty packets, discarded lengths of net and rope. Not so different from the discards you'd find around any active fishing dock. And this close to major concentrations of seabirds, surely finding their way into the birds' bodies or wrapped around them.

Along the edge of the beach, a bustling market with vendors and small restaurants, a local with a bucket of fish, flanked by Peruvian pelicans, inviting tourists to feed the birds from his bucket, for a fee. I boarded one of the tourist boats with a group of our guests and headed out for a close-up look at the islands.

The small group of Ballestas Islands are part of a larger island ecosystem in the Humboldt Current, adjacent to the Paracas National Reserve on the southern coast of Peru. Seabirds like the protection from predators that islands provide, and this particular location is within striking distance of a huge biomass of anchovies. The islands have been home to millions of seabirds for millennia and the harvest of their waste, or guano, is an important contributor to the Peruvian economy. They are often referred to as the Guano Islands, a reflection of their value.

All seabird colonies are rank with the stench of huge volumes of waste that is produced over the span of a season. I remember visiting my parents during a break from Gull Island, sitting in the family room with my father when my mother came home. She walked through the front door and, unaware that I was there—exclaimed "What is that stink?" My father called down the hall, suggesting maybe it was the dog, but my mother wasn't having it. "The dog never smelled this bad!" It is a smell that permeates everything.

The Ballestas Islands are no different in this respect. But because the region is so arid, the fetid waste does not wash away in cleansing rains. Instead, the guano of millions of seabirds has accumulated over thousands of years, resulting in huge concrete-hard mounds that were once more than seventy metres deep in some spots. These massive stores of guano are mined for their rich nitrogen and phosphorus content and used by farmers to enrich the country's otherwise poor and arid soils. In this way, the nutrients of the ocean are transported to land via seabirds.

Mining guano is the dusty, arduous and back-breaking work of pick and shovel. It's an ancient practice, at one time involving kidnappings from the South Pacific and slave labour. Damage to the seabirds or the breeding islands was punishable by death. *Guano* is an Inca word, meaning droppings. Not surprisingly, the seabird of greatest value to the industry bears the same name—the Guanay cormorant, followed by the Peruvian booby. These two species make up over 90 percent of Peru's seabird population. There is no longer a death penalty for damage to the islands, but harvesting is still done by hand to protect the seabirds that guano mining depends on. Historical overharvesting has led to the drastic reduction of guano to just a couple of metres in most places.

I had chatted with a biologist working in the Paracas National Reserve while waiting to board the tour boat destined for the Ballestas Islands. He told me that Peru had incredible resources but admitted with some embarrassment that they hadn't always taken good care of them. Today guano is harvested at a much reduced, sustainable rate of about twenty thousand tons a year and is carefully managed through the Guano Islands, Isles and Capes National Reserves System.

The air and the ocean were bursting with life as we approached. The sight of so many birds concentrated in one spot was a familiar one, but the species were completely exotic to me. Guanay cormorants, breeding in the thousands; Peruvian boobies and pelicans. The accumulated guano blanketing the arid islands was a mere suggestion of the historical volumes that once existed on these and other islands and headlands, crowded and spectacular. Inca terns, unlike any other—a deep sooty grey, with cranberry bill and matching legs, an extravagant plume of white extending from the base of the bill and curling like a handlebar moustache. Partly hidden by the moustache a bright-yellow disk of flesh. Our small boat entered a cave and flushed a colony of terns from their nests, screaming in complaint. Humboldt penguins, lined up and staring out to sea, in quiet contrast to the racket of thousands of noisy greetings and crashing waves. The bark and growl of fur seals and the high-pitched yip of pups. The revving of engines, tour boats changing course to avoid getting dashed against the cliffs. As a former seabird reserve manager, I was dubious about the boat's activity. As a newcomer, I was enthralled.

The ship continued farther north still, where the warm waters of the eastern tropical Pacific meet the Humboldt Current, off the coast between Peru and Ecuador. Like other areas of ocean convergence, it is a zone of immense richness. We were treated to thousands of Franklin's gulls in breeding plumage, migrating en masse to their colonies farther north to raise their young. Spinner and spotted dolphins streamed past the ship as far as the eye could see, churning up the ocean surface like a storm. Occasionally hitching a ride on the bow wave before peeling off to join the rest. Chased by shearwaters, hoping for an easy meal. Sprinkled amongst them, a long list of other species. Blue-footed boobies looked much like their Peruvian relatives until their legs extended during a dive, revealing the bright waxy-blue feet, a colour that shouldn't exist in nature—the colour of a hydrangea. Tens of thousands of seabirds and dolphins migrating and gorging. The spectacle lasted all day, and with it I experienced a deep sense of relief that the natural world was still managing to thrive here, despite everything. That shadow of dread that so often dogs me when I ponder the fate of our oceans had lifted, if only briefly.

Three days later, the news reported a mass stranding of dolphins on the coast of Peru—one of a series that had seen the deaths of over three thousand dolphins that had washed up along that section of the coast. Peruvian biologist Carlos Yaipen of the Scientific Organization for Conservation of Aquatic Animals attributed the mortalities to seismic surveys that were being conducted as part of oil and gas exploration activities in the waters nearby, to characterize the ocean floor. The surveys work by shooting an air gun at the ocean floor and measuring the time it takes for the sound wave energy to bounce back to an array of receivers. The blast is so powerful, the sound can penetrate hundreds of kilometres into the seabed. In the water, they can be heard four thousand kilometres away. Seismic surveys create the loudest man-made sounds in the ocean, with the exception of explosions. They kill plankton and fish. The effects on marine mammals include everything from interfering with communication, even at great distances, to death, if the animals are too close to the sound source. No one knows what that distance is.

I thought again about the biologist I had met at the dock a few days earlier and what he'd said about Peru's rich resources that had not been well cared for. With thousands of dolphins washing ashore, I had to agree with him. He had no more cause to feel embarrassed than most, though. It is the same story the world over. I need look no farther than the once-epic fishing grounds of the Grand Banks of Newfoundland, where seabirds and whales still gather to feed. The recent approval of the Bay du Nord oil project means seismic surveys will be ramping up. The threat of oil spills will remain for the life of the project. It is not only Leach's storm-petrels that are at stake, but all marine life and the fisheries that depend on those waters.

The ship left the Humboldt Current behind and entered the warm tropical waters of the Pacific. The tone of the activity changed accordingly, settling into a slower, more relaxed vibe, acquiescing to the demands of the hot, sun-drenched tropics. In these warm waters of sparse marine life, the wildlife acquiesced too. Gone, the hustle and bustle. It's the kind of place meant for seabirds that can take their time, soaring with little need for effort. The place of magnificent frigatebirds. In the early mornings I temporarily traded in my binoculars for a yoga mat, leading an easy class on the outside deck. Lying on my back and opening my eyes, I saw my first group, hanging in the air, following the ship. With their two-meter long narrow wings, forked tail and narrow, hooked bill—reminiscent of a teenaged boy—all arms and legs. Frigatebirds can fly over the ocean for months at a time, almost never landing on the water—a seabird with no waterproofing. Its gangly appearance belies its adept manoeuvrability. It chases other seabirds until they drop their food, then sweeps in and grabs it mid-air. Or it will catch a flying fish, grazing the ocean surface. All without getting a feather wet. It sleeps with only half its brain for ten seconds at a time, the other half alert. A grand total of forty minutes of half-brain sleep a day while at sea. I have raised three children and feel like I've lived this for short periods of time. I can't imagine picking it as a strategy for life.

The lazy pace continued through the Panama Canal, to Costa Rica and the protected islands of the Caribbean, bathed in the warm waters of the Atlantic's Gulf Stream. Most of the seabirds of the Pacific did not cross the land bridge to the Atlantic, but some did, including the red-footed and brown boobies. Audubon's shearwater appeared—an Atlantic species and the only shearwater found in the Caribbean. *Sargassum* seaweed everywhere.

The ship docked in the port city of Puerto Limón in Costa Rica and the expedition team had a few hours off to explore on our own. I had heard that there was a small boat available that would take you to Uvita Island, a small island not far from port. It was not a tourist destination and the boat that would take you there was hard to find. I joined a couple of friends from the team to seek it out, and was happy for the company. We hired a Jeep that took us well off the main drag, bouncing down dusty dirt roads, pockmarked with potholes, until we came to a house on the edge of a lagoon. The Jeep driver exchanged a few words in Spanish with the guy on the dock and we were beckoned to board a small boat in broken English, more gestures than language. The boat slowly wound its way through the lagoon, and we were privy to a sight not captured in the Costa Rica tourism brochures: Dense foliage on either side, alive with the calls of monkeys and birds. Thick vines hanging just above the dark, still water. The shoreline choked with garbage, a wattled jacana with its long toes walking over the water surface on floating plastic bags, a flash of bright yellow under its wings. The boat parted the floating debris, gently pushing it toward the shore in its wake. I did a visual inventory of what would have littered the lagoon if not for plastic: one bicycle frame.

For days we visited remote Caribbean islands. Before any landing, there is a briefing on board the ship, to inform guests about what they can expect and how they should prepare. Before our first island visit, our assistant expedition leader, Helga, raised the issue of cultural sensitivity. Helga is an anti-plastic-pollution advocate and had spoken of plastic pollution and Hurtigruten's commitment to reducing its own plastic footprint several times during the trip. So her message was in some ways surprising: "Please do not remove litter from the beach. People live here. This is their home and we are their guests. You will see litter scattered ashore, but please remember that it is not our place to impose our values and we do not want to insult their hospitality."

Helga was right. In places, the white sand sparkled with the foil of chip bags and other snacks. Towels under shade palms competed with weathered plastic bottles and plastic labels from water and soft drinks, peeled off the original bottle. The islands were small, the beach backshore giving rise to palm trees, shrubbery, and in the insular community—with no garbage management system—the trash had become integral to the landscape. Like an invasive species that had taken hold and flourished.

After seven weeks on board the ship, I had a few days off. From Miami, I leapfrogged most of the eastern seaboard, rejoining the ship in New York before continuing north. We visited picturesque towns in Nova Scotia and Quebec and took in the northern gannet colony on Île Bonaventure in the Gulf of St. Lawrence—the new feeding range of the North Atlantic right whale. I kept my eye out for them, knowing that they would be arriving any day. Nervously taking note of every rogue wave that hit the ship (a wave or a whale?) and monitoring the ship's speed with a sideways glance at the display on the main deck (maintained, always, below ten knots). Then onward to the colder Labrador Current, where seabirds were returning by the millions to begin their summer breeding season, and where floating buoys marked fishing gear.

One thing was clear from this long journey: marine plastic is everywhere. Washing up on beaches, or carelessly discarded there. Fishing nets and lines spilling from docks, or dragging behind active fishing vessels. The plastic in evidence, like the tip of an iceberg—a sign of what lurks below the surface, unseen and poorly understood. The dead dolphins that happened to wash up onshore were a horrific reminder of that. How many others died but did not wash ashore? What else died with the dolphins as a result of our pursuit of more oil? There is a tenuous relationship between human activity and the survival of marine life that shares the same waters. We are asking the ocean to support our insatiable desire for "more." But the more we ask of the ocean, the less it has to give.

Chapter 10

The Raven's Parachute:
Plastic and the Gulf Stream

When I was in elementary school, our class would occasionally go skating at the local arena. One of our favourite games involved forming a long human chain that spun in a circle, anchored by the person at one end, standing at the centre of the ice. That person would start to slowly spin, setting everyone else in motion. The farther out the chain you were, the faster you spun, and that's where the biggest thrill was. The chain broke up only when the outermost person would peel off, no longer able to hold on, or too scared to keep it up. The rotation of the earth works in a similar way—the speed of the rotation depending on where on the planet you are measuring it. The closer to the equator, the slower the rotation. All things being equal, by the time you get to the poles, you're the last kid in the chain.

At least in this very limited way, ocean currents behave like children. And while the equatorial waters of the Gulf Stream set a languid pace for our cruise through the Caribbean in 2019, the current picks up the pace as it veers north, carrying with it the heat absorbed from the atmosphere near the equator, and pulling with it anything in its sway, from *Sargassum* to plastic.

The pace and force of the Gulf Stream isn't as straightforward as a simple response to the speed of the earth's rotation; the contrast with surrounding water temperatures plays an important role as well. As the warm waters of the Gulf Stream displace cold northern waters, sending them deeper and directing them southward, the Gulf Stream loses heat and becomes known as the North Atlantic Current. The exchange of warm and cold waters influenced by the Gulf Stream has a major impact on climate.

Think of trying to warm a room in winter. You close the door to the hall and crank up the heat—only to find a cold draft blasting through the bottom of the door. The dense, cold air is rushing in to fill the void left by the heated air, which rises as it warms. If the temperature were the same on both sides of the door, there would be no draft. There is a similar dynamic at work between contrasting water temperatures and densities.

Global ocean temperatures have been rising over the last hundred years. As oceans warm, the temperature contrast with the Gulf Stream will diminish and slow the system—and all the other interrelated ocean circulation patterns that they drive. Oceanographic surveys both reveal and monitor these changes.

In 2018, I conducted seabird surveys aboard the CCGS *Hudson* on one of DFO's routine oceanography research cruises—a rare chance to venture into the dynamic Gulf Stream. Their research serves to characterize the ocean and to monitor changes over time in the main physical properties of sea water: conductivity (salinity), temperature and depth (CTD). Oceanographers take these measurements by lowering an electronic sensor (known by the same initials) into the water column at designated sampling stations. The CTD is housed within a rosette of bottles that can be "fired" (opened to collect water samples) at assigned depths. In this way, water samples can be collected and analyzed for chemical and biological content at these various depths. The CTD and rosette are spooled out over the ship on a kilometres-long cable that can reach the ocean floor.

It is important to revisit the same stations each year, but weather and other factors can—and usually do—affect sampling plans. Although the coastal waters of Nova Scotia are influenced by the Gulf Stream, the edge of the continental shelf is where the current really starts to exert itself. Here, the water depth drops off, from about 50 metres to 5,000 metres at the bottom of the slope. The core sampling program attempts to cover this whole depth range each year.

Ocean temperatures changed considerably, warming as the ship travelled off the edge of the continental shelf and over the continental slope into deeper water. This is where the warm waters of the Gulf Current, deflected by the continental shelf, start to have an influence. The changes in water temperature were reflected by the marine life I was seeing on the surface, the colder water species giving way to those typical of warmer regions—great shearwater to Cory's shearwater—and the exotic appearance of flying fish and Portuguese man o' war jellyfish. The warming air that accompanied them offered a welcome change for us heat-starved Canadians after a long winter.

Usually on this mission, there is a mere dip of the ship's toe into these subtropical conditions before heading back to the continental shelf and Halifax harbour. But the chief scientist, Dr. Igor Yashayaev, was a physical oceanographer—the branch of oceanography devoted to the study of physical properties of the ocean, including currents. He had a special interest in the Gulf Stream. The core program had been completed with some time to spare, a rare set of circumstances. It was an extravagance of time and energy, but Dr. Yashayaev couldn't resist the opportunity to go further and sample into the heart of the Gulf Stream. And so the ship continued southeast into waters it hadn't worked in for a very long time. A place I had never been. Finally, I could understand just how powerful the Gulf Stream really is.

When the ship entered the Gulf Stream, the wind picked up dramatically—from southerly 10 knots to gale force 35 to 40 knots. The power of the current was dragging the ship off course, and because the warm water was not allowing the engines to cool, the ship's speed was greatly reduced. Even the colour of the water was different, taking on a deep-blue hue, lacking any green that would suggest a plankton bloom. Several attempts were made to lower the CTD into the water column, but the ship could not be held steady on station. Rather than sinking, the sensor was being swept along the surface with the current. Efforts to sample had to be aborted and we turned back to the northwest, toward the continental shelf and the home port of Halifax. As soon as the ship exited the Gulf Stream, the winds dropped off dramatically again and changed direction to north 10 knots. Being on the continental shelf felt like being on the bank of a raging river. And no wonder: according to NOAA, the Gulf Stream moves close to four billion cubic feet of water per second, an amount that exceeds the volume of all the rivers in the world combined.

A map of the global distribution of marine plastic accompanies a scientific paper that addresses the issue. It clearly shows that the force of the Gulf Stream reaches into the High Arctic and carries plastic with it. Each white pixel on the map represents twenty kilograms of floating marine plastic. As you might expect, the five gyres, where plastic is known to be highly concentrated, are a solid white. From the North Atlantic Gyre, the course of the Gulf Stream can be traced in the chalky line of white pixels that traces a path to the High Arctic along the Gulf Stream's northern branch, the North Atlantic Current. Between Norway and Greenland, a fog of white pixels—a fog of marine plastic.

———

There are three main areas where large volumes of water enter and leave the circumpolar Arctic Ocean: the Davis Strait, the Bering Strait between Alaska and Siberia to the west, and the Fram Strait between Greenland and Norway to the east. The North Atlantic Current provides by far the greatest input of all currents moving into the Arctic Ocean, and it enters through the Fram Strait, feeding into a complex system of warm and cold water circulation through the entire Arctic Ocean. The relatively warm waters of the North Atlantic Current have huge implications for marine productivity in the Arctic and, as we've seen, for the distribution of plastic in what would otherwise be a relatively pristine environment.

Strictly speaking, the Arctic Ocean is the deep ocean basin around the North Pole that is permanently ice-covered, and it is encircled by the shallower waters of the continental shelves, which are ice-covered seasonally. It is these peripheral ocean waters that support marine life, with its dynamic annual changes. As Arctic ice diminishes, so too does the opportunity for these animals to survive.

Cold surface waters are generated the same way in both the Arctic and Antarctic regions. While the cold deep-bottom currents are formed by the creation of sea ice, the cold surface currents are created from its melting, along with glacial and river runoff.

Because salt gets expelled from sea water as it freezes, sea ice is actually made of fresh water, or pretty close to it. Icebergs, formed from the compression of thousands of years of snowfall, are also fresh. So in spring and summer, when all of this ice melts, there is a huge release of fresh water into the polar oceans. What results is a clash between cold fresh water at the surface, warm salty water and upwellings of cold deep-bottom waters—all with different temperatures and densities, butting into islands and continental land masses and spiralling off, forming gyres. It's an energetic and complex system that creates what is known as the Arctic Ocean Circulation. And as we've seen, ocean mixing creates fertile conditions for marine life to flourish.

The sea ice that forms each year is critical to marine life in both the Arctic and the Antarctic. The salt that gets pushed out during sea ice formation carves small channels as it sinks to the bottom of the ice, creating a briny slush. This super-salty solution supports an entire ice ecosystem of salt-loving micro-organisms that can thrive in relatively poor light conditions, including some species of algae known as ice algae. The ice provides a perfect substrate for the algae to grow on, and unlike older, thicker ice, light can penetrate it. Algae begins to grow in the fall as the ice forms, but goes dormant through the dark winter. When the light returns in spring, the algae blooms rapidly, and a dense carpet grows on the underside of the ice. The algae is grazed on by zooplankton, including one of the most important species of the Arctic food web, the tiny, fat-rich copepod *Calanus glacialis*. Females rise from the ocean depths and feed voraciously on the growing ice algae in April, fattening up for egg production. When the sea ice melts a few months later, the nutrients and organisms held in the ice are released. This, coupled with the extensive daylight of polar summer, leads to a massive plankton bloom, and all the grazing zooplankton gorge themselves on the abundance. By this time, the baby copepods are old enough to feed on the free-floating phytoplankton feast. But the gorger also becomes the gorgee.

These fatty and nutritious nuggets form the base of the Arctic food web, being the primary food for Arctic cod, bowhead whales and dovekies. Other whales, seals and seabirds feed either on the copepods and other zooplankton directly or on the Arctic cod. *Calanus glacialis* is not the only species of zooplankton in the Arctic. But with a body that can be up to 70 percent fat, it is the prime rib of its marine ecosystem.

In 2018, I sailed on the *Amundsen*, a Canadian Coast Guard research vessel and icebreaker that travels to the Canadian Arctic each summer with university and DFO research scientists and students studying aspects of marine ecology and oceanography. As usual, I was there to conduct seabird surveys and to keep track of any marine mammals I saw during my efforts. We passed Prince Leopold Island, one of the most important seabird breeding colonies in the Arctic for a multitude of species, including tens of thousands of northern fulmars, black-legged kittiwakes and several thousand black guillemots. The spectacular vertical cliffs that rise 250 metres above the sea provide breeding habitat for 100,000 pairs of thick-billed murres. The murres rely heavily on the ice-associated Arctic cod, and their breeding season is timed to when the sea ice—and therefore the fish—-are within close range of the colony. By mid-August the island was mostly abandoned for the season, the white-washed cliffs the only evidence of the earlier abundance. Guillemots and fulmars swam along the edge of the remaining pans of ice, taking advantage as the ship crushed and folded the ice, exposing small prey that had been safely hidden beneath.

None of the abundance there would be possible without the copepods and other zooplankton. And that food base would not exist without the complex interplay of ice and ocean and currents. Sea ice is essential to Arctic life, and warming oceans means less of it.

Increases in melting Arctic ice have been detected in the Bering and Fram Straits, with potentially devastating consequences. During a 2015 mission through the Davis Strait, we travelled through expansive ice fields. When I returned five years later, there was no sea ice to navigate and only a very few icebergs. An unusual year? Maybe. But I was reminded of a polar bear and her cub I had seen in 2018, swimming fifty kilometres off the coast of northern Labrador. The exhausted cub repeatedly climbed up on its mother's back to take a break, her body sinking a little deeper. The cub would eventually slide off, swimming a bit farther before needing another rest. This pattern repeated until they disappeared from sight. I looked nervously around the ship on all sides, searching through my binoculars for ice, a chance for mother and cub to get a reprieve. There was none. In my correspondence with polar bear scientist Ian Stirling, I mentioned this incident and asked for his professional opinion on their likely fate. From what he could gather, there was hope for the mother to reach land, but he wasn't optimistic about the young cub.

I have been lucky enough to travel through the Northwest Passage several times. In addition to the research trip aboard the *Amundsen*, I have been part of the expedition team on several cruises through the Passage. During those trips I hadn't seen much that would suggest a marine plastic problem in the High Arctic. Not on the surface and not from a ship, at least.

In 2021, I was part of a team that conducted aerial surveys in some of the areas that I had sailed in the past, and the view from the air offered an entirely new perspective. Normally, the flow of ocean currents cannot be seen, but the rivers that flow from glacial rivers carry silt with them and illuminate complex patterns, like chalk drawings. Intricate interweaving strands suggesting a spiderweb or the veins of a leaf, or billowing cumulus clouds in shades of white, mauve and dusty blue. The sharp line of a front like a torn sheet of paper, where two bodies of water met but refused to mix. Or a long plume, wending, serpentine, near the coast. I also saw sheets of thick blue insulating foam in Eclipse Sound, the occasional plastic bag, unidentifiable industrial-sized, bright-red discs—I couldn't guess what those were for. Was there more plastic here, or was it simply that I was seeing more from three hundred metres above the water?

I asked one of my colleagues, Jayco Tatatuipik, an Inuk in his fifties from the community of Arctic Bay, if he had noticed a change in plastic debris in the ocean over his lifetime. He said that he has seen a big difference, starting in the 1980s, particularly along the coast. He didn't attempt to quantify; instead, he told me the story of a raven he had seen flying, with one wing through each handle of a plastic bag. The white bag dragging behind, inflated like a parachute.

That image of the raven still troubles me. Normally so agile and deft in the Arctic sky, now stuck with the burden of lugging that bag. Like the raven, the ocean pulls plastic around with its currents, wherever they go. The raven and the ocean, both imperilled by it.

———

The lingering influence of the Gulf Stream also explains the delightfully temperate climate of northern Europe, the last gasp of its warmth exerting its influence beyond any reasonable expectation. Like the balmy conditions I encountered on the north coast of Denmark in the late summer of 2020.

We were preparing for another oceanographic research trip in the Davis Strait, which runs between Greenland and Baffin Island in the Canadian Arctic. This time it was on the Danish ship *Dana*, with a collaborative research team from the United States, Denmark, Greenland and Canada. The chief scientist, Dr. Craig Lee, arranged the very complicated logistics. He needed to find a port country that would allow relatively high-risk Americans to travel there in the midst of the COVID-19 pandemic. Denmark was the place. We were to leave from the port town of Hirtshals near the country's northwest tip.

At the eleventh hour, the Canadian oceanographers were denied permission to travel to Denmark due to COVID-19 restrictions. I was the lone Canadian and joined Dr. Lee and the rest of the American team in Denmark for travel and quarantine within the country before boarding the ship.

We headed to Løkken in northern Denmark, located at 57° north latitude, exactly halfway between the latitudes of my hometown of St. John's, Newfoundland, and the Arctic Circle. On the Canadian side, 57° north latitude puts you in Okak, Labrador—a hundred kilometres north of Nain, Labrador's most northerly community. A region of subarctic tundra, perfectly suited for polar bears and Arctic fox. I did not pack shorts for Denmark.

There were no Arctic fox in Løkken, the heart of a family resort region, with sandy beaches that stretch as far as the eye can see. Families on holiday drove onto the beach and set up for a day of

building elaborate sandcastles, playing some beach version of cro-
quet, sunbathing, swimming. Comfortably. The backshore was all
sand dunes covered in long, sweeping beach grass. Behind that were
summer homes, hobby farms, shade trees, deer. This is what the Gulf
Stream will do for you. I bought a bathing suit.

It was a nine-day commute from the Danish port town of Hirtshals
to the Davis Strait—more time than necessary or desirable for the
oceanographers, eager to get to work in the Davis Strait and the
Labrador Sea. I would be conducting seabird surveys uninterrupted
for days on end, in areas I'd never been before, in waters where
dynamic forces were brought to bear. First up were the shallow
waters of the North Sea, infused with the North Atlantic Current
(the north branch of the Gulf Stream). We then weaved between
Shetland and Fair Isle before heading into the deep open ocean
south of Iceland, where the confluence of warm and cold currents
from the Arctic and the Gulf Stream negotiated their terms, coming
to various arrangements at different depths, all the way to the ocean
floor. Exchanging heat and salt—the currents altered, the names
changed. Then it was back into more familiar waters, skirting along
the edge of the Labrador Sea before climbing onto the continental
shelf of Greenland. Here, the remnants of the Gulf Stream's warmth
push northward along the southwest coast of Greenland, forcing
that northward migration of sea ice that feeds the Labrador Current.

The opportunity to cut across this swath of the swirling, mixing
subarctic currents was new for me. I was interested to see how these
complex and interconnected ocean currents shared and exchanged
waters from such varied origins, and what the sum of those parts
would add up to, reflected by the life we could see at the surface.

As the ocean bathymetry changed, so did the wildlife. There were fewer black-backed gulls, black-legged kittiwakes and northern gannets in shallow coastal waters; the scant offshore sightings of the tiny European storm-petrel disappeared completely at the edge of the continental shelf. Between Fair Isle and the Shetland Islands, it was surprisingly quiet. But then again, it was September. These islands, so frantic with seabird life during the breeding season, had been abandoned for this year, the birds' presence as ephemeral as the availability of shoaling fish they rely on to feed their young.

I was joined on the bridge by Dr. Kate Stafford, a physical oceanographer and whale research scientist who was part of the American research team from the University of Washington. This was a new area for Kate as well. We were both eager to get to the edge of the continental shelf, the area of upwelling and mixing that we knew held promise for more activity. The Faroe and adjoining Bill Bailey Bank just west of the Faroe Islands did not disappoint. The edge of the banks dropped off steeply to water depths of 1,500 to 3,000 metres. Manx shearwaters were a constant even in deep water, but overall the seabird numbers were not particularly dramatic. The whales were another story. The combination of shelf and deep water gave rise to minke, fin whales, pilot and sperm whales, Risso's dolphin. We travelled through deep water for the next two days. There were so many whales I had to stop recording the usual narrative in my notes; it was too time-consuming and compromised seabird observing. South of Iceland we hit whale pay dirt—seven different species of whales, including three separate sightings of blue whales.

The bridge was crowded with enthusiastic oceanographers, and Kate provided a running commentary on what we were seeing, sharing aloud her thought process until arriving at a positive identification:

"large blow . . . at the surface a long time . . . it's going to dive . . . see the ridges on the tail stock . . . flukes . . . sperm whale."

The fervour on the bridge was palpable—the high-pitched tones, the laughter, the chorus of *whooooas* at another spectacular encounter. I was trying, with varying degrees of success, to focus my attention on birds, standing off in my corner. For me, spectacular encounters like these have become attended with an undercurrent of unease in recent years. These enormous creatures, so adeptly adapted to their marine environment, but so vulnerable to the barrage of novel threats they are helpless against—ship traffic, seismic activity, fishing gear, plastic garbage, climate change. Normally, I can't separate pleasure in the experience of seeing marine wildlife from dread of the forces working against their survival. Are they healthy? Are they safe? Are they finding food? But there was so much life bursting at the surface around us that day. Once again, I was filled with a sense of optimism. Despite the unrelenting global environmental angst, there was still burgeoning life here.

Then, a routine scan to the port side and Kate spotted a pinprick of bright orange, out of place on the horizon. One hundred nautical miles due south of Iceland, it could well be the rounded hull of a lifeboat. She brought it to the captain's attention and he changed course to get a better look. As the ship drew closer and the object took form, it was clear that it was not a lifeboat. It was a large buoy, adrift. Used for fishing gear, maybe a trawl line. We couldn't tell. Satisfied that there were no human lives at risk, the captain turned the ship around and headed back on course. And I imagined kilometres of hooked line below the surface, waiting.

Chapter 11

The Plastisphere: Microplastics, the Oceans and Human Health

What do ocean currents have to do with marine plastic? Everything. I have seen the plastics that are filling the oceans, carried impassively by the same currents that have distributed oxygen and nutrients around the globe since the beginning of time, swept into this global circulation of the planet's life-giving resources. Spewed out from choking rivers and along coastlines, swirled by currents, carried to the deepest and most remote regions of the world's oceans, and threatening the creatures that live in them. The ship cuts a narrow line through the ocean, a hair's breadth. What I have witnessed in the swell and chop: a plastic bag here; a block of misshapen Styrofoam with pieces missing there; a water bottle, a sheet of floating plastic; ghost fishing gear; faded-orange, weathered, broken, unidentifiable objects enmeshed in seaweed hundreds of kilometres offshore—and this is only a hint of what lies beneath the surface.

Knowing that it is out there, the next question is: What is it doing?

MICROPLASTICS

Before diving into the research, I used to think that fishing gear was the most dangerous plastic for marine life. But I was wrong about that. Plastic presents different threats, depending on its size and texture, whether it is at the surface, in the water column or on the ocean floor. And the greatest danger is posed by the plastic you can't even see.

Like naturally occurring marine detritus, microplastics eventually sink. In a recent study, water samples were taken in the Great Pacific Garbage Patch at depths of up to a thousand metres. They found that the plastic smog is actually denser at depth than at the surface.

Some plastics are micro by design. Microbeads are the tiny, hard, spherical balls of plastic that are used in an array of personal care and hygiene products, including cosmetics, body cleansers and toothpaste—products that are designed to be washed down the drain. I have been addicted to facial scrubs since they first came on the market and never showered without them. Then someone posted an article on social media about microbeads, with a status that imparted the caution: microbeads are wreaking havoc on the marine environment, so don't use products containing them. *What?*

Is that even possible? What harm could these little particles possibly do? When it comes to awareness of microplastic pollution, I was late to the party. I read the article, albeit reluctantly (did I really want to know this?), and switched to apricot scrub, no biggie.

It wasn't long before another post cautioned: your fleece and other synthetic clothing is shedding plastic microfibres into the ocean every time you wash them. What fresh hell is this? Almost everything I own is fleece or some stretchy, comfy, unnatural Lycra-containing material. Essentially, anything that is not made from a natural fibre like cotton, linen, wool or hemp is made of synthetic plastics (remember polyester?). Almost every article of my clothing was partially or completely plastic. All contributing to marine pollution, one microfibre at a time. If you consider the contents of your dryer trap, you get some idea of how much shedding your clothes do. Unlike in the dryer, the microfibres shed in the washer go straight down the drain.

Microplastics, at five millimetres or less in size, can be invisible to the naked eye—about the size of a sesame seed or smaller. Primary microplastics are called pre-production pellets, or "nurdles"; tiny by design, they form what is essentially the raw material for the manufacturing of plastic products. Nurdles are themselves a source of microplastic pollution, escaping into the environment at every step of their development and transportation—from pellet production, packaging and shipping to product manufacturing at far-reaching locations. Nurdles have been found on beaches around the world.

Large plastic materials that break up into small pieces are considered secondary microplastics. Whatever the source, these plastic particles can be carried in the air and float freely in the ocean. Lighter than naturally occurring detritus, they can travel farther in air and water currents. Microplastics have been found everywhere, from near the summit of Mount Everest to the guts of small crustaceans dwelling in the Mariana Trench, over ten kilometres below the ocean surface. Seventeen different sources of plastic were found in a 2019 study of Arctic ice cores.

A recent study conducted by DFO sampled microplastics across the Arctic, from just below the surface to a depth of a thousand metres. Scientists identified the microparticles using Fourier transform infrared spectrometry (FTIR), a technology that can identify the polymer types of the tiny particles. They found forty microplastic particles per cubic metre of water, 92 percent of which were microfibres. Almost three-quarters of the microfibres were polyester, pointing to synthetic clothing and textiles as the source.

The study also found that there were three times more particles in the eastern Arctic (above the Atlantic Ocean) than the west (above Alaska and the Yukon) and that the eastern particles were longer and less weathered. This suggests that the majority of the plastics entering the Arctic originated from Europe, eastern Canada and the United States, then moved northwest, becoming degraded along the way. This makes sense, since the majority of the ocean currents flowing into the Arctic originate from the Fram Strait above northern Europe, in the Atlantic. A more recent study has found evidence that the majority of the plastic originates from European rivers.

Microplastic is everywhere, and there's plenty more to come. It is so ubiquitous, the term *plastisphere* has been coined to describe the community of organisms that grow on it. Just as the biosphere is the thin film of life on the surface of the planet, the plastisphere is the microscopic film (biofilm) of single-celled bacteria, algae and other microbial life that readily grows on the surface of microplastic. There are communities of thousands of species that thrive on even a single piece. A couple of concerning issues here: Some of the bacteria are pathogens, carrying disease that is transmittable to humans and marine wildlife. And because microplastic can travel great distances with the currents, it has the potential to introduce invasive species to other areas. The associated microbes may in turn disturb the natural ecosystem of the new region they have entered and have implications along the food web.

The ratio of size to surface area is much greater for small objects than it is for large ones. This means that if you were to break up a block of plastic, the surface area of all the small pieces would greatly exceed that of the original block. Microplastic particles therefore have a large surface area relative to larger plastic pieces, and as plastic breaks up, the surface area increases, which allows for greater leaching of toxins from plastic into the environment. It also provides a larger surface area for absorbing toxins from the environment, including PCBs and heavy metals. For their weight, microplastics pack a more toxic punch.

Microplastic floats through the water column like naturally occurring suspended particles and can be mistaken for food by tiny predators (zooplankton). Several lab studies have looked at plastic consumption by copepods and have shown that copepods will "graze" on microplastics even in the presence of algae, their natural food. In one study that used polystyrene, copepods consumed up to 40 percent less real food, choosing the microplastic instead. As a result, the copepod itself is not getting the nutrition it needs because the plastic also tricks the body into thinking it is being fed, circumventing the natural mechanism to slow metabolism during periods of low food availability, to prevent starvation. As microplastic density grows in the ocean, it can be expected to have an increasing impact on the health of copepods and other zooplankton—as well as their nutritional value as prey for other species. Dovekie diets depend on copepods. A study that examined the gut contents of dovekies from a group that washed ashore in Newfoundland found that of the 171 birds sampled, 30 percent contained plastic. The plastic was thought to have been either consumed directly or contained within the prey they ate. Of the dovekies that were studied from the Newfoundland beach event, 99 percent of the plastic was postconsumer waste. Seventy-eight percent of that was polyethylene, followed by polypropylene (21 percent). Their tiny prey are becoming, quite literally, junk food. North Atlantic right whales depend on copepods as their main food source too. As plastic works its way up the food chain, the substance accumulates, along with whatever toxins are associated with it.

In areas like the Great Pacific Garbage Patch, where microplastics form an ocean "smog," they are downright impossible to avoid. With the exception of zooplankton, most plastic ingestion occurs randomly and quite by accident. In the case of microplastics, no matter the concentration, you can't avoid what you can't see.

A great deal of plastic is eaten incidentally by animals while feeding on other prey. In the case of baleen whales that feed by gulping huge mouthfuls of small schooling fish, plastic of various sizes in the water column can be ingested during a feeding frenzy, along with the fish they are targeting. An autopsy of a humpback whale revealed the presence of a range of plastics in its digestive tract, including polyethylene (milk jugs, shampoo bottles, plastic bags, six-pack rings), polypropylene (straws, cups, fishing rope), polyvinylchloride (cling wrap, plastic pipes), polyethylene terephthalate (the biggest single-use polymer in use, for example in water bottles and fast-food containers) and nylon (clothing, rope, auto parts and tires). The plastics found in the gut also ranged in shape and size, from one millimetre to seventeen centimetres.

Microplastics are consumed by everything from the smallest zooplankton to the largest whales. They are eaten by fish and siphoned out of the water column by filter-feeding shellfish like mussels, oysters and scallops. They can also enter the bodies of fish and crabs through their gills. Together, the ingredients of a delicious seafood paella. What affects the ocean's health directly affects our own.

HUMAN HEALTH

Long thought to be inert, plastic is now known to leach toxins into the environment, particularly when stressed by exposure to heat, sun, weathering and other chemicals—and to soak up toxins from the environment like a sponge.

The scope of the effects of plastic and its additives on our bodies is not well understood, but growing evidence of human health concerns has spawned further research in the field. Links have been made to cancers in men and women, and to metabolic disorders and problems with sexual development, particularly in fetuses and young children. And that's just for starters. But studying plastic effects is tricky. As with any scientific study, to prove an effect you need a control group for comparison—a group that has not been exposed, in this case to plastic. Because plastic is found everywhere, there is quite literally no one who has not been exposed; no one is eligible to form a control group. Still, startling impacts on humans have been documented, with profound physiological consequences.

To date, the greatest concerns come from the additives bisphenol A (BPA), phthalates and brominated flame retardants. Plastic resins containing BPA are hard and clear, giving the appearance and some of the advantages of glass but with the added advantage of being lightweight and less breakable. BPAs have been widely used in reusable water bottles, baby bottles and food can liners, to name just a few examples. Phthalates are a component of flexible plastics like polyvinyl chloride (PVC), which is employed in an array of products, from pool liners to food storage cling wraps. Other forms of phthalates stabilize fragrance in lotions and various personal care products. Brominated flame retardants (BFRs) are another group of chemicals, which are commonly used to reduce the flammability of an array of materials, including household furniture, mattresses and textiles.

It is now well-established that these chemicals leach out of plastics through the processes of weathering, heating and aging—and have become wide-ranging contaminants of natural ecosystems, as well as of tissues in the human body. BPA and phthalates are pervasive pollutants, and particles have made their way into all environments, including the polar regions. BPAs are released into the atmosphere when plastics are burned, for example, and absorbed into food and water from the plastic packaging they are contained in. Phthalates are found in the air we breathe, particularly indoors. Humans are exposed to them through inhalation, ingestion and absorption through the skin from lotions, fragrances and other scented personal care products.

Their harmful effects on human bodies are manifested primarily through the endocrine system.

The endocrine system is the collection of glands in the body that produce hormones and act as the communication highway to the brain, regulating many of the body's functions, including metabolism, growth, sexual development, reproduction, sleep and mood. Endocrine disrupters are chemicals that either mimic or block natural hormone production and can wreak havoc on the body's normal, healthy functions. One group, known as estrogenic chemicals ("estrogen-like" endocrine disrupters associated with the hormone estrogen), has been linked to problems with reproduction, obesity, aggression and growth rates, to name just a few. Estrogen is found in both males and females, but is dominant in females and is involved in the development of the reproductive system and female sexual characteristics, including breast development. According to the

National Institute of Environmental Health Sciences (NIEHS), endocrine disrupters may pose the greatest risk during prenatal and early postnatal development when organ and neural systems are forming. The NIEHS is also supporting research to determine if exposure to endocrine disrupters increases the incidence of infertility and certain types of cancer.

The endocrine-disrupting action of bisphenol A was discovered by accident in 1998, while geneticist Dr. Patricia Hunt at Washington State University was conducting genetic experiments on mice in her lab. BPA plastics are ideal for many industrial and consumer products, such as safety glasses, DVDs, food containers, protective coatings for the hulls of ships—and Dr. Hunt's mouse cages. At the end of her experiment she tested her "control" mice that had not been subjected to any experimental manipulation, expecting them to be normal. To her great surprise, she discovered that 40 percent of the control animals had egg defects. After months of investigation, she concluded that a cleaning product used to wipe down the cages had scarred the hard plastic surface, enabling the chemical BPA to leach out. The BPA had severely disrupted normal estrogen function in the mice. This raised alarm bells for potential effects on human health.

In 2010, BPA was declared a toxic substance by Health Canada and taken off the market for food applications. But other countries have been slower to respond and the bans have been limited. Partial bans were imposed in the US starting in 2012, when the FDA amended its regulations in response to a petition brought forward by the American Chemistry Council (ACC). The new regulations banned the use of BPA resins in baby bottles and sippy cups. The following year, infant formula packaging was added to the list in response to another petition, this one filed by Congressman Edward

Markey in Massachusetts. However, in 2018 the FDA stated that BPA did not represent a health hazard when used in food containers, and the ACC announced that an FDA study had concluded that current uses of BPA are safe. The problem with this, according to Dr. Hunt, is that the FDA uses an antiquated standard to assess safety: whether it is toxic enough to cause death. Dr. Hunt says the standard should reflect whether it causes health problems. I'm with her. But even at their current standard, the FDA may want to reconsider their position on the use of BPA. The review paper points to lethal and sublethal effects of both BPA and phthalates. Sure, you may survive breast cancer after a gruelling course of chemotherapy, radiation and possible mastectomy—but wouldn't it be better to avoid it in the first place?

In 2011, the European Council of Justice ruled that BPA is toxic for reproduction and a substance of very high concern (SVHC)—a special designation with legal implications. As in the US, the resulting ban was limited to baby bottles, but a ban similar to Canada's is expected soon. The lobby group PlasticsEurope challenged the decision, hoping to make it impossible for BPA to be listed as an SVHC. This designation would oblige the plastic industry to share information with the public about the chemical's hazardous effects and how to mitigate them. However, the General Court of the EU upheld the decision.

Even where bans on BPA have been implemented, the chemicals that replace it are not necessarily an improvement. When Dr. Hunt looked at bisphenol S (BPS), a common replacement for BPA, she found it caused genetic aberrations in egg and sperm production, resulting in a reduction in viable sperm and an increase in egg abnormalities. Further studies have revealed almost exactly the same results for other replacement chemicals—BPF, BPAF and diphenyl sulfone. If the chromosomal abnormalities caused by low doses of these replacement chemicals in rodents caused the same problems in humans (as was the case with BPA), then even if bisphenol contaminants were completely eliminated from the environment, these problems would persist for about three generations. As it turns out, "BPA-free" is a red herring.

In July 2020, the journal *Birth Defects Research* published a comprehensive overview of the endocrine-disrupting actions of BPA and the phthalate family of plastics. In it, Dr. Phillipa Darbe summarized the known impacts of these chemicals on the human endocrine system. Here is what she found:

BPA mimics estrogen, and it has been shown that in vitro exposure increases the expression of genes that are estrogen-sensitive. It has also been shown to increase the enzyme aromatase—the key enzyme required for the conversion of androgens ("male" hormones) to estrogens ("female" hormones). In addition to mimicking estrogen, BPA can also bind to androgen receptors, creating antiandrogenic responses. Not surprisingly, it has been shown in animal studies to negatively affect male and female reproductive functions. Finally, BPA can bind to thyroid receptors and impact thyroid hormone production—the hormones that regulate metabolism and affect every organ in the body.

The potential impacts on the body from this hormone-mimicking compound are profound, and include not only sexual reproductive and thyroid functions but also immune function, diabetes, cardiovascular disease and obesity. The review goes on to say that exposure to BPA has been linked to hormone-sensitive cancers, particularly breast cancer.

I recently ran into an old high school friend and we did some catching up on each other's lives. She is a practising physician with a particular interest in community health. It wasn't long before the subject turned to plastic. When she was thirty-six, she was diagnosed with breast cancer. At the time, she delved into the research to try to make sense of her diagnosis, particularly at such a young age.

The scientific literature is couched in conservative language— "may cause," "is linked to," "could be." I have used it myself—I certainly can't go further than the reported findings do. But my physician friend had no such constraints when it came to managing her own disease. Even twenty-two years ago, the medical literature convinced her to eliminate plastic from her family life as much as possible. Her kids complained about the weight of their lunch boxes, with their embarrassing—and heavy—glass containers for sandwiches and snacks. I reflected on what I sent my own kids off to school with—the plastic baggies full of carrots and grapes. I thought I was packing a healthy lunch.

One report in particular stood out for her. Many breast cancers are driven by estrogen. To establish whether a cancer is one of the estrogen-driven types, biopsied breast tissue is tested for the presence of estrogen hormone receptors. To this end, a reagent is added to both breast tissue samples and a control group (samples that do not contain breast tumour cells) to ensure that the reagent is working properly. The authors reported a puzzling result: all the samples, including the control group, indicated the presence of estrogen receptors. Upon further investigation, they discovered that the estrogen-like hormone was leaking out of the plastic receptacles and contaminating the results.

She concluded, in exasperation, that over the years, plastic food packaging has become increasingly difficult to avoid. Today, it is almost impossible.

Now, phthalates.

Like BPA, some phthalates are known to mimic estrogen, causing similar concerns. The literature does suggest one novel impact, though: in utero exposure to phthalates in rodents has been shown to negatively impact male reproductive development, creating a condition that is "strikingly similar" to testicular dysgenesis syndrome in men. Testicular dysgenesis syndrome includes an array of factors related to male reproductive health, including poor semen quality and a heightened risk of testicular cancer. The disorder has become increasingly common at the same time that fertility rates are decreasing in the Western world, and is considered a major health problem. There is consensus in the literature that exposure to phthalates strongly suggests a significant negative impact on reproduction in men. However, the authors of a scientific review published in 2021 in the *American Journal of Public Health* go much further in their position. In light of all the mounting evidence, the authors call for immediate policy reforms to eliminate ortho-phthalates from any products that might lead to exposure in women who are pregnant or of childbearing age, as well as infants and young children.

Polyethylene terephthalate (abbreviated to PET or PETE) is one of the plastic polymers that was created by Dupont in the 1940s during the post-bakelite rush to find and develop new plastics for commercial use. PET is a clear strong plastic and one of the most common plastic polymers, widely used for packaging foods and beverages. Virtually all single-use water bottles and soft drink bottles are made of PET. If you look around your own house at personal and household care products that are contained in clear plastic bottles, you will find that they are made of PET. I have done it in my own home to confirm: Clean & Clear face wash, Johnson's baby oil, Bath & Body Works liquid hand soap, Lush toothpaste tabs, Dove body mist, Benylin cough syrup, Kirkland fish oil capsules, Parker & Bailey wood floor cleaner—the list is endless.

When the material is used for fabric, it is known as polyester. Over half the world's synthetic fibre is made from PET. These are the microfibres we are washing out of our clothes and into the ocean. My beloved fleeces are made from PET.

Because PET is so commonly used and so abundant in the environment, it deserves special scrutiny. According to the PET Resin Association (PETRA) website in March 2022, polyethylene terephthalate doesn't contain phthalates. Of course it does—it's right there in the name. The distinction they are attempting to make, if slightly overstated, is that the ortho-phthalates used in plasticizers are derived from ortho-phthalic acid, while PET is made using tere-phthalic acid. PETRA is distancing itself from the ortho-phthalate group that are already known to be toxic. But how reassured should we be?

PETRA states that "drinking water from a PET bottle that has been left in a hot car, frozen, used more than once, or repeatedly washed and rinsed does not pose any health risk." They go on to say that "PET is a very inert material that is resistant to attack by microorganisms, and does not react with food products . . ." Yet recent scientific findings call these claims into question.

Although all of us inevitably consume plastic, the amount taken in by bottled water drinkers far exceeds that of everyone else. A review paper published in 2019 looked at the results of twenty-six studies on human plastic ingestion and found that, based on recommended daily water intake, those who drank only tap water consumed about 4,500 microplastic particles a year. Those who drank an equivalent volume of only bottled water consumed an *additional* 90,000 microplastics annually. The same review of plastic ingestion found that, on average, adults consume at least 50,000 microplastic particles a year, children about 40,000. *Levels were drastically higher in people who drank bottled water.* The researchers examined both marine and terrestrial sources of plastic, including tap water, bottled water, beer, fish, salt and sugar.

Another paper, published in late 2021, reports a link between the concentration of ingested plastic in stool samples and inflammatory bowel disease. By far the most common microplastic was PET. I would call that a health risk. So is PET really as safe as PETRA claims? I don't know. The problem, for me at least, is that all plastics have been considered safe until it turned out they weren't.

PETRA also asserts that PET contains no endocrine disrupters—or at least no *known* endocrine disrupters. I don't doubt this statement to be true, but given the problems with previously approved plastics that have now been linked to serious human health problems, I remain skeptical about the presence of *unknown* endocrine disrupters.

Some endocrine disrupting chemicals, have become known as obesogens, because they lead to fat accumulation in the body. Studies have shown that exposure to BPA can interfere not only with energy metabolism but with the very structure of body fat, by increasing the number of fat cells, or the amount of fat stored in the cellsleading to excessive weight gain in children and obesity in adults. Studies across multiple generations in rodents have shown that, following exposure to BPA and some phthalates, these obesity traits can also be inherited by future generations and can dysregulate hormonal control over hunger and satiety.

Prenatal and early postnatal exposure to BPA and phthalates has also been implicated in altered brain and immune system development, and in cognitive and behavioural problems in children. This is pretty unsettling, to be sure. But it gets worse. A recent study that examined the placentas of six people with normal pregnancies found microplastic fragments in four of the six placentas that were examined. Twelve pieces were found in all, ranging in size from five to ten microns (0.005 to 0.01 millimetres). Five were on the infant side of the placenta, four on the maternal side and three in the membrane between the two. At these minuscule sizes, microplastics are referred to as nanoplastics, and are capable of crossing cell membranes.

Of the twelve pieces, only three could be positively identified. All three were polypropylene. To refresh your memory—this is the same polymer used to make plastic straws and rope used extensively in fishing gear. This should not be surprising. Microplastic is an equal-opportunity pollutant, and plastic straws are not just a problem for turtles. As for polypropylene ropes—lost and discarded fishing gear makes up the majority of the macroplastics in the ocean, and about 10 percent of the global ocean's microplastic. Calculated from the European Parliament 2018 report, that amounts to about 12.3 million metric tons of microplastic.

Of the remaining nine nanoplastics found in the placentas, only the pigments could be identified. Their uses, listed by the authors, include man-made coatings, paints, adhesives, plasters, children's finger paints, cosmetics and personal care products—products with potential for both BPA and phthalate contamination.

A recent literature review examined the reported in vitro effects of plastics on other species. The data showed that through oxidative stress and suppression of acetylcholinesterase (an enzyme vital for communicating nerve impulses throughout the body), micro- and nanoplastics can potentially contribute to the development of neurological disorders. As with all studies on the impact of plastics on human health, the authors conclude that more research is urgently needed to assess the neurotoxic risk caused by exposure to microplastics.

An article published in the *American Journal of Public Health* in 2021 is unequivocal about its findings, determining that ortho-phthalates—found in such products as food packaging, medical supplies, cosmetic and other personal care products—have been shown to leach into food from the plastic tubing used in commercial dairy operations, food preparation gloves and food packaging materials, to name a few. The authors have determined that exposure to ortho-phthalates can impair children's brain development in a number of ways, including an increased chance of developing ADHD and other behavioural issues, and can also cause adverse cognitive development, resulting in lower IQ. As a result, the authors call for critical policy reforms that would eliminate ortho-phthalates from products that lead to risky exposure for infants and small children, including exposure for pregnant people, and even people of reproductive age.

Finally, we come to the brominated flame retardants. There are over seventy-five different BFRs, but one of the most common is tetrabromobisphenol A (TBBPA), found in essentially all environments around the world. TBBPA is generally excreted quickly from the body. Even so, it has been detected in human plasma and breast milk. Recent studies have linked BFRs to endocrine-disrupting activities affecting the function of the thyroid, reproductive and immune systems. According to a 2021 UNEP report, flame retardants associated with plastics have been implicated in neurodevelopmental disorders (including attention deficit hyperactivity disorder, or ADHD, autism and cognition), thyroid disease, thyroid cancer and decreased antibody response to vaccines.

Polystyrene foam (commonly known as "styrofoam") is a plastic that does not contain BPA or phthalates. Industrial forms may include BFRs, but food containers do not. To clarify, there are actually two types of polystyrene foam: Extruded polystyrene (XPS) and expanded polystyrene (EPS). The extruded version is harder, denser and carries the trademarked name Styrofoam--the "real" Styrofoam. The expanded foam results from a different process, creating a lighter product, the type of polystyrene foam that is used for food containers (as well as a multitude of other things) and has become commonly known by the same name. All this to say that not all "styrofoam" is the same. So when you read about BFRs and other toxins being added to Styrofoam, it refers soley to industrial applications. It is the polystyrene itself that deserves closer scrutiny when it comes to our food containers.

Polystyrene is a polymer formed through chemical processes that use benzene as the building block to make styrenepolystyrene and finally polystyrene foam. Benzene is a known carcinogen. It off-gases or leaches out, particularly when it is heated. A serious problem for products designed for hot food and beverages. When ingested, benzene causes damage to the lungs, nervous system and reproductive organs. In a word, polystyrene foam is toxic. Unfortunately, another word that describes the product is *convenient*. Polystyrene foam is cheap, lightweight and an effective insulator, keeping the heat in your food and away from your skin. Do the benefits outweigh the costs? As with fossil fuels, the benefits of plastic products and compounds are obvious and immediate. The costs are often hidden, or don't reveal themselves for years or even decades.

The levels at which plastic compounds become toxic to humans are difficult to prove unequivocally. Dosing rodents to measure toxic effects is one thing, but for obvious reasons, it is both unethical and illegal to perform dosing experiments on humans. Still, we can see that many of the experimental results found in lab rodents have been manifested in human populations. The cause of these issues is surely related to many different factors, but just as surely, the contribution of plastic cannot be discounted. In the face of the mounting scientific evidence, why would you risk it?

There is one last point that is worth thinking about, as we consider our dependence on plastic products and the throwaway mindset. Those products that are not created for holding or storing food and beverages are not evaluated for their level of toxicity to humans. BPA, phthalates and all other toxic compounds are still widely used for commercial, industrial and domestic applications—the assumption being that you are not going to eat your bleach bottle, for example. But when these products are tossed into landfills or dumped into the environment, the toxic chemicals leach out and contaminate groundwater, streams and eventually the ocean. Similarly, when they break up into microplastics and move freely through the air and water, they do not discriminate. There are no regulations governing where they end up—either on your skin, in your lungs or on your dinner plate. Talk about unexpected consequences.

Four

Petrochemicals and
the Future of Our Seas

Chapter 12

Whales on Toast

The petroleum industry is primarily responsible for climate change and solely responsible for the plastic crisis we find ourselves in. As we move toward finding solutions to plastic pollution, there are some important lessons to be learned from the early days of the oil industry. Or, to be more precise, before petroleum, when whale oil was the world's fuel source.

South Georgia, located in the rich waters of the Antarctic Convergence in the Southern Ocean, has an infamous history of marine exploitation. The first sealers arrived in 1788 in pursuit of fur seal pelts, and in less than fifty years over a million fur seals had been harvested. By 1912, the seal industry ended, the species perilously close to disappearing altogether. At about the same time as the sealing industry was sputtering out, the whaling industry in the region was ramping up.

At that time, there was a huge demand for whale oil, primarily for use in lamps and for making candles. The first whaling station to service the Antarctic industry was established by the Norwegian C.A. Larsen in 1904 at Grytviken, South Georgia. The slaughter of the large whales—blue, fin, sei, humpback, right and sperm whales—was relentless and lasted for sixty years. By 1965, when the last station closed, 175,250 whales had been processed in South Georgia, and the large whales had been driven to the brink of extinction.

In what seems like a mystifying development, the demand for whale oil grew in the first half of the twentieth century, despite the availability of an alternative fuel. Environmental sociologist Richard York explains how the discovery of petroleum did not replace the demand for whale oil, but actually led to an increase in hunting pressure, threatening the species' very survival.

The fossil fuel industry started in the mid-1800s and was driven in the early days by the invention of the kerosene lamp and the oil derived from coal that was needed to fuel it. Petroleum was a much cheaper and more viable option than whale oil; and with the creation of the kerosene lamp, whale oil was no longer necessary for lighting. The emergence of petroleum production should have offered a much-needed reprieve for the whales, but relief was very short-lived. Whale oil remained the primary driver for the whale harvesting industry throughout its history, long after petroleum was in wide use. So why couldn't whales cut a break?

As it turns out, the relationship between the petroleum and whaling industries is more nuanced and complicated than you might imagine. It is well worth having a closer look, because the driving force behind the unrelenting slaughter was our all too human desire for more—more oil, more consumer products, more profit. And, maybe not surprisingly, it has direct implications for how we move forward as we try to combat both climate change and plastic pollution. If we pay attention to history (not our strong suit), it is clear that simply finding alternatives to plastic will not crush the demand for it any more than the discovery of oil decreased the demand for whale oil.

The early days of commercial whaling date back to the Basques in the eleventh century, using the technologies of sailing ship, small open rowboats and hand-driven harpoons. The ships transported the open boats to within reach of the hunting grounds; the boats were then launched either from shore or from the ship, and the oarsmen attempted to chase down the whales. They had to get close enough to allow for a hand-thrown harpoon to make its mark, finishing the kill with spears. The carcass then had to be rowed back to either ship or shore. Not only was the early commercial hunt gruelling and danger-ous, but the technology was slow. Men could not out-row most whale species and thus were limited to the large, slow-moving family of right whales. As the name implies, they were the "right whale" to hunt because they were slow, lingered at the surface and had the added bonus of floating once they were killed—essential for safely towing the animal from the site of the kill. For the time being, the other large baleen whales, which were much faster swimmers, were spared.

By the sixteenth century the industry had greatly expanded, involving several European countries. The technology was still fairly primitive, but the hunt was ruthlessly successful. Thousands of right whales and bowhead whales (their buoyant brethren) were killed in the North Atlantic and the Gulf of St. Lawrence in this way.

The whaling industry went through some major innovations, made possible by fossil fuels. First came the coal-fired steamships, which were much faster than those under sail, enabling whalers to pursue the faster, large whales. Soon after, Svend Foyn, a Norwegian whaler and philanthropist, invented the harpoon cannon. It featured an explosive tip, which could kill even blue whales, the largest on earth. These harpoons could be mounted on the bow of the ship, which eliminated the need for rowboats altogether; you had to like that if you were an oarsman. Finally, the development of fossil-fuel-driven air compressors created a means to pump carcasses with air to keep them afloat. Now, essentially all whales were the "right whale" to hunt. And thus began the modern era of the whale hunt.

Paradoxically, the modern whaling technologies could not have been developed without the ready availability of fossil fuels. But why was it necessary, or even desirable, to pursue whales when petroleum was pouring out of the ground? The answer is contained within a mix of greed, ego, patriotism and market manipulation.

Assuming that one oil source will simply replace another is to assume that there is a fixed demand for oil; it ignores the all-too-human drive for *more*. Other fuel alternatives besides kerosene were already available, including plant and domestic animal fats in the form of lard and tallow for making soap, candles and lubricant. Still, this was a time of population and economic growth. The volume of oil from petroleum production and the falling price of whale oil did not kill the whaling industry, but actually intensified it from the 1860s to the turn of the twentieth century. The harder something is to get, the more we want it. Enter supply and demand.

For example, right whales and sperm whales of the North Atlantic and Pacific were severely depleted by the 1860s, which limited the profit potential from hunting them. But because whale oil had become hard to get, its value—though falling—hadn't plummeted from its peak in the 1850s, despite the competition from petroleum.

If whale oil still had value, then the next logical consideration was how to get more when there was admittedly very little left.

The answer lay in technology. Steam and diesel engines allowed for faster commutes from whaling stations in the Southern Ocean to markets in the north. At about the same time as whaling extended to the south, there was another game-changing innovation: factory ships enabled whales to be processed and the product to be stored on board. These ocean-going factories were free to roam the world for months at a time, supported by whaling ships that supplied them with carcasses, returning to port only when their storage capacity was filled. The industry was further aided by the invention of the freezer, a fuel-based technology that allowed for the long-term storage of meat. By the 1920s, factory ships were common.

In summary, petroleum fuelled the development and expansion of technologies that accelerated the slaughter of more whales—and more species of whale—than had ever been achieved before, leading to a surplus of whale oil on the market. Would this cause a crash in whale oil prices? You would think so. Now enter the 1989 film *Field of Dreams*, and the disembodied voice in Kevin Costner's head that promised, "If you build it, they will come."

The turn of the twentieth century was a period of great innovation, with rapidly developing novel products and global markets. A great example is the development of hydrogenation, a process that allowed whale oil to be a solid at room temperature and removed the unpleasant odour, creating whale margarine. As Richard York dryly puts it: "Thus in the 20th century, whales were killed in huge numbers so that they could be eaten on toast . . . Consumers were convinced they wanted whale margarine, even in places where there was plenty of locally available butter."

If you are amused by this early consumer gullibility, don't laugh too hard. Consumer manipulation is alive and well. We may not be spreading whales on toast today, but we are spending a fortune on bottled water, even in places where an ample clean and free source is flowing from our taps.

The writing was on the wall for whales if harvest was allowed to continue at an unmitigated rate. In 1946, the International Whaling Commission (IWC) was created to manage and conserve global whale populations while attempting to support a sustainable harvest, but quotas were lax and the measures proved ineffective. So much so that whalers gave up on the hunt in the Southern Ocean in the 1960s because the dwindling numbers made the long journey unprofitable. By 1982, with many species showing only very slow recovery, if any, the IWC called for a moratorium on commercial harvesting. It went into effect in 1986.

There were countries that refused to honour the moratorium to varying extents and for slightly different reasons, mostly boiling down to shades of patriotism. The Soviets engaged in political posturing, defying international quotas during the Cold War. To this day, Japan, Norway and Iceland defy international pressure to cease their hunts, motivated by values related to cultural heritage and identity.

Still, the vast majority of countries respect the moratorium on whale hunting, and populations of most species are slowly recovering. It can be argued, then, that whales were spared extinction by legal protection and not by the petroleum industry, which, in theory, could have taken pressure off the resource. There is absolutely no doubt that modern whaling hastened the journey to the brink of extinction, but I also think it's pretty clear that there would have been no appetite for protecting whales if there wasn't another fuel option readily available.

If we follow York's line of reasoning, there is another logical conclusion to be drawn about the future of the plastic industry: Climate change action is driving the development of alternatives to petroleum for fuel. So what does the petroleum industry do? Create and expand petrochemical plastic products, in order to buoy the petroleum industry.

As more sustainable energy sources are being developed to combat climate change, oil producers are shifting their gaze from fuel toward the plastic industry to bump up the demand for their product. Plastic production relies on oil production; the International Energy Agency (IEA) projects that the production of plastic and other petroleum products (petrochemicals) will drive oil demand heading toward 2050. But the direct link between plastic production and greenhouse gas emissions seems to have been left out of the conversation. In a 2018 interview with Reuters news agency, the IEA executive director Fatih Birol said, "The petrochemical sector is one of the blind spots of the global energy debate and there is no question that it will be the key driver of oil demand growth for many years to come." Plastic contributes almost a billion metric tons of greenhouse gas emissions each year, and if we continue on this path, those emissions will triple by 2050.

In response to the anticipated demand, oil companies such as ExxonMobil and Royal Dutch Shell are investing in new petrochemical plants, as are some countries in the Middle East, including Saudi Arabia and Kuwait, knowing that they can make more money converting oil to plastic rather than products like gas and diesel. Saudi Aramco, for example, announced in 2018 that it plans to invest over

$100 billion in petrochemicals over the ensuing decade, expanding in Saudi Arabia as well as other overseas markets in China and India. Petrochemicals are expected to increase the demand for oil by 7 million barrels a day by 2050, for a total of 20 million barrels a day. But plans don't end in 2050. According to Saudi Aramco's president and CEO Amin H. Nasser, the ultimate target is 8 to 10 million barrels a day—and that is just one company.

If we have learned anything from the wanton slaughter of whales, then we know that the solution to the marine plastic crisis has to be tackled on several fronts. Finding viable alternatives to plastic is critical, but it will not eliminate the demand for new plastic unless there is a simultaneous change in policy and regulations that supports healthy oceans—and a deepened understanding of what is at stake. There is an urgent need for a plastics version of the International Whaling Commission, and an international body to regulate plastics, if we want to reverse the damage wrought by decades of environmental neglect. We are not there yet.

Chapter 13

Out of Sight, Out of Mind

During the few days I had off during the long trip from Antarctica to Newfoundland in 2019, I met up with my husband in Manhattan—visiting museums, strolling through Central Park, wandering the streets and eating out. Normal tourist stuff. But the unbridled abundance I'd witnessed from Antarctica to the Americas was alive in my mind; I couldn't stop thinking about how the addition or subtraction of something invisible or apparently inconsequential could change everything—ocean current, water temperature, bird guano, plankton, plastic trash. From the unblemished purity of Antarctica to the strobing glitz of Times Square, I was seeing Manhattan through a different lens than my husband.

A person can spend weeks in this port city without ever seeing the ocean. Skyscrapers. Central Park, with its sprawling lawns, trees, paths, vendors, a boathouse and restaurant: an urban refuge. Street performers at its edges. And people everywhere—in the park, on the streets. Tourists gawking, locals with places to be, a more urgent gait. At lunchtime, the restaurants are packed and efficient, set up for take-out and quick turnover. Fast-food joints: five minutes of eating, then dump your overflowing tray of coated paper, waxed or plastic cups, and Styrofoam shells into the garbage receptacle. More upscale spots: artisanal breads loaded with all the on-trend healthy ingredients, kombucha or green tea or decaf latte, double sugar, hold the fat. Still, the single-use cup, the single-use clamshell, the single-use cutlery, the single-use straw. Maybe eco-friendly-looking brown paper napkins. All tossed in the garbage or, covered with food scraps, into recycle bins. Which, then, magically disappear. Downtown Manhattan—you could eat off the streets. Where does all the garbage go?

Sure, there are dumpsters in back alleys full to overflowing with garbage. Out of sight, out of mind. With an efficient waste management system, the city's garbage literally disappears from view. It's understandable that most people don't concern themselves with it or wonder about where the garbage or their dirty recyclable materials are taken. By dropping their waste into the appropriate bin, they've done their duty. On to the next board meeting.

What if people were made aware that their dirty plastic waste is being packed onto container ships, bound for countries in Central America and Southeast Asia without the infrastructure or regulations to deal with it? Their plastic bottle of pure mineral water, now swirling in the tropical waters of the Andaman Sea, along with hundreds of tons of other plastic garbage dumped there—floating on the surface, suspended in the water column, filling the open mouths of manta rays. Or lies on the ocean floor, smothering coral reefs. And those balloons, sent into the air in a grand gesture of love or celebration? Any idea where they went, or what they did when they got there? I had some idea, having witnessed the floating armada of Mylar balloons off the coast of New York City on that Valentine's Day in 2017. The loggerhead turtle in slow but steady pursuit. But most people never see this outcome.

When it comes to our refuse, most people behave in a way they believe to be responsible and expect that things will be taken care of as they should be.

The advent of the COVID-19 pandemic has given new life to this "throwaway living" mentality and to the plastics industry. A desire for the sterile and the safe has led to a major regression in our willingness to reuse just about anything.

In 2021, I was hired to work with an aerial survey team in the Arctic. The team was housed at a mining camp on Baffin Island. Due to COVID-19 restrictions, all personnel were required to take food from the cafeteria, in single-use plastic containers, to their rooms upon arrival at camp. For the first eight days I saved the containers, hoping I would find a use for them. I hadn't accumulated plastic waste like this since my field season in Sunnyside in 2016, and it made me feel a little bit sick. I invented uses for some of it, but most piled up, stacked in layers and flooding over my desk, inevitably ending up in the garbage. Baggies, clamshells of every size and shape, bananas and other fruit wrapped in cling wrap, all served in a plastic bag. After the eight days you were allowed to use dishes and eat in the cafeteria, but food was often served up in disposable plastic containers anyway. If you wanted reusable tableware, you had to make a point of it. Most people didn't. Plastic drink bottles were collected for recycling, and all other plastic was incinerated on-site, great plumes of smoke rising into the Arctic sky.

Restaurants and food services all over the world moved toward single-use plastic as an immediate solution to concerns about contamination and the spread of COVID-19. Disposable items like masks proliferated. In 2020 an estimated 129 billion face masks and 69 billion disposable gloves were used globally *every month*; about 2 billion face masks ended up in the ocean that year. Perhaps because we are so accustomed to the single-use mindset, a more sustainable solution has not been considered for the long term, as the pandemic drags on.

Since the 1960s, the medical establishment has employed disposable products to limit the spread of bacteria. But before then, reusable latex gloves were commonplace in the operating theatre. These gloves could be sterilized and were made to last. A surgeon would retain his pair over an entire career, moulded perfectly to his hand and cared for like the violin of a concertmaster.

Disposable latex gloves replaced the reusable gloves; reusable cotton gauze masks were replaced by paper and then synthetics. By the end of the 1960s, most hospitals were moving toward a system of complete disposability. Single-use garments had the health benefit of guaranteed sterility, along with the convenience of being able to toss them rather than wash them. As with the emergency measures taken for COVID-19, there was no apparent consideration given to the long-term consequences, or whether there might be a better way to protect human health while factoring in environmental impact.

During the early days of the pandemic, when the personal protective equipment (PPE) supply chain was under strain, health care professionals were reusing masks that had been effectively sterilized to address the shortage. As soon as single-use disposable became an option again, there was an immediate halt to this practice. Although reuse was short-lived, it was a reminder that there are viable sterile and safe alternatives to complete disposability. Ones that we are nowhere near ready to embrace. Yet.

Any time I enter a hospital or medical clinic, I am issued a fresh disposable face mask. Recently I asked what happened to all the rejected masks we were required to wear before we got to this checkpoint, pointing at the garbage bucket near-filled with them. Are they recycled? The attendant responded that, as far as he knew, they were incinerated. Outside the hospital entrance, a ballistics-splatter of the freshly issued disposable masks, blowing away from their source. And like other discarded plastics, headed toward streams, rivers and, almost certainly, the ocean.

The COVID-19 pandemic aside, we have known for some time that plastic has been accumulating around the globe to untenable levels, but efforts to address the problem have failed so far and plastics continue to accrue. New plastics are being churned out at the rate of 380 million metric tons each year. In order to move forward, we have to first look back—to understand why we find ourselves in a plastic crisis and why efforts so far have failed to curb it.

The Möbius loop was launched on the first Earth Day, in 1970, and is recognized today as the international symbol for recycling—synonymous with the adage Reduce, Reuse, Recycle. The name *Möbius* might not be familiar, but the symbol is: the three broad arrows, folded over and chasing each other in a triangle. The very notion that recycling exists has led most of us to skip the first two Rs and jump from Use to Recycle, entertaining the well-meaning but misguided assumption that the plastic will be kept out of landfills and in circulation, environmental damage averted.

The Möbius loop was never patented. In 1988, the Society of

Plastics Industry developed a Resin Identification Code (RIC) made up of three chasing arrows with a number in the centre. All plastic resins have an RIC. The number is shorthand for the type of polymer that the plastic item is made from; it has nothing to do with recycling. But the striking (and suspicious) similarity to the Möbius loop has created that impression. Polyethylene terephthalate (PETE or PET) is represented by the number 1 and, as we've seen, is used for a multitude of single uses, including food packaging and beverage bottles. High-density polyethylene (HDPE) is represented by the number 2 and is frequently used for products like milk jugs, shampoo bottles and other products with a similar texture. Low-density polyethylene (LDPE) is symbolized by the number 4 and used for plastic bags; polyvinyl chloride (PVC) by the number 3, which is used in some cling wraps.

The confusion between the Möbius loop and the RIC symbol serves both plastic producers and product packaging designers, who have not been above taking advantage of it. To see how this works, let's have a closer look at cling wrap.

Dow Chemical developed polyvinylidene chloride (PVDC) during World War II for military use. It was a thin film, used to protect sensitive equipment from moisture, oil and other potential contaminants. In 1952 it became a commercial product, available to the public under the trade name Saran Wrap. Cling wraps continue to be made of PVDC, polyvinyl chloride (PVC) or polyethylene (PE) today.

I haven't used cling wrap since I became aware of the plastic problem, a good five years ago, and recently pulled it out of my kitchen drawer to have a closer look. So I can attest to its durability—it was as good as new after all that time. Not to pick on this product in particular, but the one in my kitchen was Glad wrap. On the box are statements that suggest environmental sustainability, such as the fact that it is BPA-free. An empty recycle symbol stands out—no number contained within, because it is not referring to the product, which is a plastic resin, but to the cardboard carton it is packaged in, which is recyclable (where facilities exist). There is no mention of the material that the cling wrap itself is made of, just that it does not contain BPA. *And that it is microwave safe.*

The packaging included an invitation to call a 1-800 number with questions, so I called. I got a recorded message, and the menu options stunned me. They referred solely to medical problems related to the use of their products: "If you have used one of our products and believe you may be having a health-related response to our product, please press 1 now to be transferred to a medical consultant available 24/7. If you are experiencing a medical emergency, please hang up and dial 911."

I was left with the impression that I would be better off eating the box than covering my food with the cling wrap.

I called back and was told that Glad cling wrap is made from 100 percent polyethylene. The claim that their product is microwave safe is a bit of stretch, so to speak. In fact, a new and promising technology relies on the very fact that microwaves melt plastics, including polyethylene. More on that later.

No matter which polymer cling wraps are made of, they are not recyclable with current commercial recycling technologies. So, the box the cling wrap is contained in is recyclable and environmentally sustainable, but the product is not. This is not at all obvious to the average consumer, and it is not meant to be. It's just one example of the widespread problem of "greenwashing": a deliberate attempt to create a false impression, using misleading information or unsubstantiated claims to suggest that something is environmentally friendly, or at least sustainable.

In addition to the confusion over whether something is recyclable or not, there is a wide range of variability in recycling capacity from one facility to the next. A plastic that has a number inside the recycling symbol may be recyclable, but only where the appropriate facilities exist.

It's also generally unclear how to prepare plastics for recycling. There is the problem of recycling contamination—for example, when materials are not properly cleaned of food residue, like a yogurt cup that has not been rinsed before it's tossed in the recycle bin. You see this kind of thing in airports all the time; ditto for public recycling bins. It is referred to as aspirational recycling—reflecting the hope that by simply throwing something into the recycling, it will end up where it should be. Unfortunately, this is rarely the case. Those glass juice bottles, water bottles and yogurt cups that get combined more often than not end up in the landfill. And one dirty yogurt cup is enough to contaminate the lot. It's true that some facilities can deal with food contamination, but it's best to assume that they cannot. Dirty recyclable plastics just end up as trash and can contaminate otherwise clean plastic in the airport recycle bin or the sorting facility, rendering it all as trash.

Last but not least is the confusion arising from terms like bioplastic, biodegradable plastic, compostable plastic. All these terms suggest that the products thus labelled break down into natural constituent parts, returning to the environment and leaving it unscathed. But most bio-based and plant-based plastics contain toxic chemicals and are not much of an improvement over the standard petrochemical plastics. Compostable plastics are generally only broken down by commercial composters with temperatures exceeding 50°C, conditions not found in garden composts and certainly not in the open ocean. An experiment that looked at the rate of breakdown of various carrier bags found that the one labelled as biodegradable was still completely intact after three years in a marine environment.

In the end, most biodegradable products do not live up to the promise their names imply. They don't break down easily and have similar problems to those associated with petrochemical plastics: absorbing toxins, breaking down into microplastics, and becoming vectors for the growth and transport of pathogenic bacteria. And a bioplastic bag-loop noose is just as lethal as any other. One question remains, though: If a marine animal ate a biodegradable or compostable bag, would it get broken down in the warm acidic environment of the gut? That could be a distinct advantage, but I have not been able to find any information on this matter.

It is fair to say that there are some serious challenges around making informed choices, even when you want to do the right thing. Lost in the ambiguity of it all, some have just given up trying. And, let's be honest, not everyone tries.

Another reason for the plastic pile-up is the ready availability of cheap virgin plastics—plastics that are derived from petrochemical feedstocks and have never been used or processed before. Of the 380 million metric tons of virgin plastic produced each year, between 35 and 45 percent is dedicated to single-use packaging—throwaway plastic, by design.

The current linear system has resulted primarily because the economic opportunities in recovering plastics have not been recognized. In the absence of economic incentive, recycling rates have been abysmally low and recycling has not turned out to be the panacea it was once thought to be.

The world's richest countries have dealt with the spiralling plastic problem by trying to make it disappear, mostly by shipping plastic garbage offshore, generally to poorer, less developed countries with waste management laws and facilities that are inadequate, at best. The sheer amount being shipped has overwhelmed the ability of these countries to recycle or repurpose the plastic, so it simply piles up, choking rivers and inundating beaches.

China was once the dumping ground for the world's recyclable plastic waste. The landscape bore the legacy: until very recently, six of the twenty most plastic-polluted rivers in the world are in China. The Yangtze river had more plastic than the top twenty polluting rivers put together. China finally called a halt to all imports in 2018, which created a huge plastic backlog, and many countries in Europe, the US and Canada subsequently tried to offload the surplus onto South American and Southeast Asian countries. Many countries on the receiving end of those shipments are awash in plastic. The river pollution statistics reflect the change. In 2021, the Pasig River in the Philippines contributed at least six times more plastic than any other river. Eight of the the top 10 most plastic-emitting rivers are found in the Philippines and Malaysia. Overwhelmed by the volumes and environmental impacts, Vietnam, Thailand and Malaysia have also taken action against foreign imports. In 2019, Malaysia demanded that countries—including Canada—take back a whopping three thousand metric tons of their plastic waste.

Some remote islands find themselves on the receiving end of plastic waste carried by the currents, like the ones we visited in the Caribbean on the expedition ship. Here, plastic is an abundant, free and "renewable" resource; it is burned as fuel for cooking. But there is a price to pay: the acrid smoke laden with toxic chemicals is released into the air and has led to respiratory diseases and other health issues among the population.

The practice of burning plastic is not limited to the remote islands of low-income countries. Incineration is one of the means used in landfills around the world for reducing garbage volumes, including a significant amount of plastic. Over 756 million metric tons of plastic has been burned in landfills globally. In 2016 the journal *Procedia Environmental Science* published a review summarizing the toxic effects of this practice on the environment and on human health. According to the article, plastic makes up 12 percent of municipal solid waste; about 40 percent of the world's garbage is burned, releasing gases into the environment including dioxin, furans and polychlorinated biphenyls (PCBs)—all chemicals that are not only highly toxic but known to be carcinogenic. According to the World Health Organization, dioxins can also cause reproductive and developmental problems, damage the immune system and

interfere with hormone function. And the list of harmful effects goes on. The chemicals released by burning polystyrene (commonly known as styrofoam) can damage the central nervous system. The brominated fire retardants act as mutagens as well as carcinogens.

The soot generated by burning plastic contains toxic components including heavy metals. Some of the soot is airborne and is carried by the wind and deposited anywhere—on land (including agricultural areas) and in both fresh and marine waters.

In this way, plastic incineration creates pathways for dangerous toxins to enter the environment and our bodies through the food we eat, the water we drink and the air we breathe. But it is not the only route. And when plastic is burned, CO_2 is released into the environment. We still struggle to find ways of dealing with plastic that do not create greenhouse gas.

Take PET, for example. You will recall that PETRA listed resistance to microbial breakdown as one of its advantages. While it is resistant, it is not immune. In 2016 a paper was published describing the discovery of a bacterium in Japan that does exactly that, albeit slowly. Experimental manipulation increased the rate of breakdown by 20 percent and further again to six times faster in 2020. Still another enzyme was found in a compost pile, reported by researchers in 2021. More on that later.

Whether PET can be broken down or not raises another important question: Which is better? When processes are applied to break down PET or any other plastic—by either bacterial digestion, incineration or one of the many other engineered processes—CO_2, as we've seen, is released into the atmosphere, because of plastic's petroleum origins. Solving the plastic problem may exacerbate the climate change problem, both of which are only going to increase if we continue on our current path.

Suffice to say, the crisis is not going to go away unless there are aggressive, immediate and coordinated international efforts and supporting legislation to cauterize the flow of virgin plastic production and the inevitable release into the environment. We need efforts like those taken to protect whales from extinction. We need a complete paradigm shift.

And there is good news here on several fronts.

Chapter 14

Reframing Plastic: The Potential of a Circular Economy

I was featured once in a magazine, complete with an extensive interview and accompanying photographs with descriptive captions. It might have been more of a flyer, really—for a local second-hand clothing store. Frenchy's was located next to my two daughters' primary school, and I would go there after dropping them off sometimes, with my preschool son in tow. One day I was browsing the "boutique" section and the store owner asked, by way of making conversation, "Are you looking for something special?" I told her I needed a dress for a formal award ceremony in Toronto, for a literary prize my husband had been shortlisted for. It required something fancy. We searched the racks together and found a two-layered number, in different shades and textures of purple—a deep, shiny layer of polyester and a breezy, flowy, paler chiffon. It fit perfectly. I promised to send pictures from the event. And yes, they proved feature-story worthy.

The concept itself is simple: find another purpose for something after its initial job is done, more or less. If you don't need it, pass it on. Or tweak it and use it for something else. It is a return to an idea that people lived by a generation or two ago: nothing that has another use is wasted or thrown out.

The septic field at our cabin was showing signs of failure and the ground above it was starting to collapse—at first visible only from the pronounced shadow cast by evening light. After a while it worsened and we started warning people not to walk there, pointing to the grassy trap. Inevitably, a friend, distracted by the story she was telling, fell thigh-deep through the grass and into the putrid filth. We had delayed dealing with the problem, mostly because of the huge cost it would entail, but a neighbour from my father's generation had a cheap, effective solution. Michael peeled back the turf of the offending area and placed the hood of an old car over it. The turf was replaced and the problem was solved. We had repurposed the car hood in much the same way the kittiwakes had repurposed the ship on Middleton Island.

As it turns out, we had been participating in the *circular economy*, as it has become known, long before I ever heard of it.

The circular economy has developed largely in response to the global plastic pollution crisis. It takes a systemic approach toward keeping resources circulating and recirculating through the economy for as long as possible, extracting the maximum value and finally recovering and regenerating materials to form new products. If you've thrifted, you've engaged in the circular economy.

There is a growing movement toward forging international plastic waste accords that promote the circular economy. One of the most ambitious is the New Plastics Economy Global Commitment, launched in 2018 and sponsored by the UNEP and the Ellen MacArthur Foundation in collaboration with the One Planet Network, the Global Partnership on Marine Litter and its Clean Seas campaign. The Global Commitment's vision is based on three basic tenets: to *eliminate*— remove plastics that we don't need; to *innovate*—ensure that the plastics we do need are manufactured such that they are reusable, recyclable or compostable; and to *circulate*—to keep all the plastics we use in the economy and out of the environment. Five hundred participating organizations will work toward a set of ambitious targets to reduce plastic waste and pollution by 2025, but this can only be achieved with all the stakeholders working together—businesses, policy-makers, scientists, governments and so on.

Let's face it, even the less ambitious goals of the reduce/reuse/recycle model haven't been effective. The phrase "take, make, waste" has been coined to describe the linear model that has actually unfolded, getting us to this dire point in the plastic crisis. However, the Global Commitment initiative is far more complex and sophisticated than "reduce, reuse, recycle," although the old adage is an essential part of the circular economy. Its vision includes a world with no negative impacts from plastic, where plastics attain their highest value along the economic value chain and no plastic leaks into the environment. The circular economy is regenerative and restorative by design. It promotes the economic opportunities that recovering petrochemical waste provides, while at the same time taking pressure off natural systems by keeping plastic out of the environment.

The initiative integrates several concepts that illustrate how a circular economy can thrive: designing out waste and pollution by developing products that do not release harmful substances in the first place; designing products for durability and reuse; remanufacturing and recycling to keep products, components and materials recycling in the economy. It requires changes in business models, changes in product and service design, manufacturing redesign—the possibilities are endless. In such a system, cellphones and other electronic waste would be broken down into their components, and the plastic would be separated out, the components that don't work would be recycled and the useful parts repurposed to create like-new products. It's exactly what your grandfather used to do in the shed, only on a global industrial scale.

Money talks and industry listens. From a global perspective, 95 percent of the material value of plastic packaging—representing $100 to $150 billion a year—is currently lost to the global economy after just a single use. Perhaps the most important element for success of a circular economy is the economic opportunities it offers. By 2024, the market value of global plastic recycling is estimated to be US$64 billion.

These principles are being implemented around the world. For example, the European Commission has adopted the Circular Economy Action Plan for a Cleaner and More Competitive Europe, which predicts an increase of the EU's GDP by 0.5 percent by 2030, creating around 700,000 new jobs. It identifies goals of 70 percent recycling of municipal waste and 80 percent of packaging materials by the same date. Some of its member states are ahead of the EU in adopting circular economy legislation.

Research and innovation incentives are also being promoted worldwide. The New Plastics Economy Innovation Prize is administered by the UK-based Ellen MacArthur Foundation and is awarded to designers, entrepreneurs, academics and scientists who offer solutions to the prevailing plastic system and the means of eliminating plastic packaging waste. Prizes have been awarded for the development of an array of bioplastic packaging solutions, using biodegradable materials such as seaweed. One innovation involves packaging single-use products from condiments to hotel shampoos in materials that dissolve in water and are completely biodegradable. There are also awards for new system designs—for example, one that offers a means of delivering groceries without the need for product packaging.

In the US, the American Bottle Association recently announced their Every Bottle Back campaign to promote the goals of a circular economy. It involves the collaboration of three major market competitors—the Coca-Cola Company, PepsiCo and Keurig Dr. Pepper—working together with the World Wildlife Fund to find solutions that keep plastics out of the environment. The campaign intends to establish a new industry fund that will expand and modernize recycling facilities and provide greater access to recycling at home. At the same time, they are developing an education and awareness campaign and working to close gaps in the circular economy. The idea is to make it easy for the bottles to be recovered and recycled to produce new bottles, reducing the amount of new plastics used in the process.

Economic opportunities in plastic recovery are central to the circular economy model. But if we have learned anything from the International Whaling Commission, it is that a good regulatory framework is necessary to achieve conservation goals. To this end, recent amendments have been made to the UN Basel Convention on the Control of Transboundary Movements of Hazardous Wastes and Their Disposal, making global trade of plastic more transparent. And making plastic pollution a crime.

The Basel Convention was created in 1989 to control the movement of hazardous waste across international boundaries in order to protect human health and the environment. With 187 member countries, the Basel Convention is the most comprehensive global environmental treaty on hazardous wastes. In 2019 it took its mandate further, adopting the Plastic Waste Amendment, which requires that certain types of plastics—those that are difficult to recycle, contaminated or otherwise hazardous—cannot be exported to another country unless the receiving country has granted a Prior Informed Consent (PIC). The PIC will only be granted if the waste is managed in an environmentally sound manner by the country receiving it. These new rules, which protect developing nations from exploitation by wealthier countries, came into effect on January 1, 2021. The amendments have strengthened the Basel Convention, making it the only global, legally binding instrument to deal specifically with plastic waste.

Canada is a signatory to the Basel Convention. However, the federal government has entered into a side agreement with the US that would allow for the trade of plastic waste between the two countries. The US has not ratified the Basel Convention and, under the terms of the agreement, has created a "pipeline" for plastic waste to be exchanged between Canada and the US.

———

Canada's response to the plastic crisis has been bewildering—as much a moving target as the plastic itself. There is light all right, but there are long shadows too. On the one hand, Canada played a leading role, as president of the G7, in developing the Ocean Plastics Charter. In it, there are lots of optimistic action verbs: *endeavour, strive, seek, encourage.* The plan does commit to taking actions toward the goals of a circular economy, through sustainable design, production and after-use markets, promoting research, innovation and new technologies, and supporting coastal cleanup. But talk with no teeth is just—well, talk. However, the Charter does reveal what we need to accomplish to clean up the mess we've made. And domestically, an intergovernmental forum has recently proposed a comprehensive strategy for tackling the plastic crisis nationwide.

The Canadian Council of Ministers of the Environment (CCME) developed the Strategy on Zero Plastic Waste, referring to data levels from 2016, and released its recommendations in two phases, in 2019 and 2020 respectively. The Phase 1 document reports that 90 percent of Canada's plastic waste is not recycled or recovered, which represented an economic loss of $7.8 billion in 2016 alone. Projecting outward, that loss is now in the tens of billions of dollars. The document also provides the context for a continued pursuit of a circular economy, reframing plastics from a waste problem to a valuable commodity—a hugely important mental shift that's necessary for economic opportunities around repurposing plastic to arise.

We know that plastic pollution has only increased since 2016, but even so, the older figures help put the problem in perspective. In 2016, Canada generated 3.3 million metric tons of plastic waste. Packaging accounted for 1.54 million tons, or 47 percent—the country's largest contributor to plastic waste. Approximately 86 percent of the plastic ended up in landfills; only 9 percent was recycled, which is in line with global recycling rates. The Strategy identified eight categories of plastic waste sources, including agriculture, and asserted that only 1 percent of Canada's plastic waste is leaking into the environment. But here's the puzzling thing: fishing gear is not mentioned, even though a background paper for the Strategy states that ghost nets are responsible for 46 to 70 percent of macroplastics in the ocean.

Back in 2009, the CCME launched the Canada-wide Action Plan for Extended Producer Responsibility, a policy meant to extend a producer's responsibility to the post-consumer stage of the product's life cycle through a framework of legislation and/or regulations, taking a harmonized approach across jurisdictions. Producer responsibility is a critical piece in successfully achieving a circular economy, but nothing substantial has happened on that front. Phase 2 of the Zero Plastic Strategy commits to addressing this shortcoming.

The Phase 2 Strategy document also commits to contributing to global action on plastic pollution reduction. To this end, the Canadian government is investing $100 million to help low- and middle-income countries develop and implement effective waste management systems. The document also outlines a framework for achieving a circular economy. Importantly, the federal government plan provides timelines for improving consumer and industry awareness, reducing pollution from fisheries and aquaculture, advancing science, supporting plastic pollution prevention and cleanup, and contributing to global action. I remain cautiously optimistic on this point, although Canada, as you will see, does not have a good track record for meeting its commitments.

The plan acknowledges the need to protect the environment from the impact of lost (ghost) fishing gear, and the government offers funding for research into this problem via the DFO's Ghost Gear Fund. The Fund supports projects that retrieve or dispose of ghost gear; develop state-of-the-art gear technology; and support international commitments. The Ghost Gear Fund will support twenty-six projects to the tune of $8.3 million from 2020 to 2022.

From a regulatory perspective, Canada has made some headway. Regulations against the import or production of toiletries containing microbeads came into force on January 1, 2018. But action beyond that has been slow. In an October 2020 news release, the federal government announced the next steps in its plan to achieve Zero Plastic Waste by 2030, describing a ban that was to be placed on "harmful single-use plastic items where there is evidence that they are found in the environment, are often not recycled and have readily available alternatives." There are enough loopholes here for a tuna to swim through. "Harmful single-use"? I think by now the case that all single-use plastic is harmful has been well-established. The question is not whether it is harmful, but rather, are we willing to absorb the environmental and health costs of its use? Medical supplies being on one end of the spectrum, Ring Pops on the other. And upon whom is the impetus placed, to provide evidence that they are found in the environment? Can't we just assume, knowing what we know, that it all ends up in the environment? There is not one type of plastic that is more likely to flow out of a landfill than another.

What's more, the list of items to be banned—the first step meant to get us to zero waste by 2030—is lacklustre at best. I would go so far as to say it is laughable, and only in the if-you-don't-laugh-you'll-cry kind of way. The items that will get Canada going on its international commitments to address the global plastic waste crisis: plastic checkout bags, straws, stir sticks, six-pack rings, cutlery and food ware made from hard-to-recycle plastics. No mention of the plastic packaging that dominates our plastic waste. Or single-use water bottles. Had anyone done the math? Even the Minister of Environment and Climate Change Canada at the time, Jonathan Wilkinson, called it a drop in the bucket. The regulations to support these measures were to be finalized by the end of 2021, but this didn't happen—the date was pushed to the end of 2022. Instead, the end of 2021 saw the announcement of a new date for reaching Zero Plastic Waste. Apparently, someone has since done the math. The Zero Plastic Waste goal was pushed back twenty years, to 2050.

Having said that, in May 2021 the Government of Canada did take an important step by revising legislation that will help deal with plastic pollution. Schedule 1 of the Canadian Environmental Protection Act (CEPA) now includes "plastic manufactured items" in the list of toxic substances. Adding plastic to the toxic substance list in and of itself doesn't do anything to affect the manufacture, supply, use or disposal of plastic, but it does provide a mechanism to regulate plastic in the future. This step is an important one because the act requires the federal government to operate in a manner that protects the environment and states that when there is a threat of serious or irreversible damage, a lack of full scientific certainty cannot be used to postpone measures to protect the environment. This effectively removes the burden of proof, which is almost impossible to establish in many situations in the natural world and is often used by industry in general to curtail restrictions on their activities (like the seismic surveys that can have such a devastating impact on marine life, as we've seen).

Having familiarized myself with the federal government's planning documents and policies around plastic pollution, I can't help noticing that they are awash with language of goodwill, and no accountability or consequences for failing to stay on track. There is a heavy price to pay for inaction. What we urgently need is strong, effective legislation.

Just as critical as strong legislation are the scientific and technological innovations that offer hope for a future beyond today's plastic crisis—and there are lots of great examples around the world right now. Scientists are sharing advancements and working together to solve the plastic problem from sometimes unexpected angles. Like bacteria, for example.

As we saw earlier, the structure of PET makes it resistant to breakdown by microbes (microscopic organisms). However, a bacterium that is capable of digesting plastic was discovered outside a Japanese bottle recycling plant in 2016—the first such bacterium ever known to science. Japanese researchers discovered that when grown on PET plastic, *Ideonella sakaiensis* could break it down with two enzymes, using its carbon as an energy source. The two by-products of the digestive process are ethylene glycol and terephthalic acid, both described as "environmentally benign" by the scientists. These by-products can easily be broken down by microbes in the environment, resulting in CO_2 and water. Yes, more CO_2.

The digestive process takes about six weeks—too long to be of any practical large-scale use. Scientists from around the world took this bacterium into the lab and started to experiment with enzyme interactions to see if they could improve the performance of the naturally occurring activity. By 2020 scientists from the US and the UK had collaborated to engineer a super-enzyme that, when paired with the two naturally occurring enzymes in *I. sakaiensis*, could break down the plastic six times faster than the natural system.

At the same time, a French company, Carbios, in collaboration with Pepsi, L'Oréal and other companies, was working with an enzyme discovered from a fungus in leaf litter that could degrade 90 percent of plastic bottles within ten hours, but the process requires heat above 70°C. The super-enzyme derived from *I. sakaiensis* has the advantage of working at room temperature.

The two teams are working together to tweak enzymes and test others to see if the process can be improved. And there are other innovations as well. Enzymes from cotton are being paired with the plastic-digesting enzymes, which could allow mixed materials and textiles (e.g., cotton-polyester blends) to be fully recycled. This is significant, since 60 percent of the world's PET production is used for the synthetic clothing fibre polyester.

Danish researcher Dr. Anne Meyer and her bioengineering team at the Technical University of Denmark (DTU) are experimenting with *I. sakaiensis* both to improve its performance and to explore new ways the by-products of its digestion of plastic can be used for other purposes. For example, the ethylene glycol that is produced can be used directly as antifreeze. The other by-product, phthalic acid, can be combined with paper to produce a shopping bag that is superior in quality to paper, and can be recycled or degraded biologically.

Meanwhile, in 2020, a German research team published their finding with a different bacteria species, a strain of soil bacterium known as *Pseudomonas* sp., which is capable of breaking down polyurethane, a plastic polymer used in a range of products from insulating foams to disposable diapers. It releases carcinogenic chemicals and other toxins as it degrades, killing most bacteria. Although this discovery is far from having any commercial application at this point, it could be very important down the road, particularly in light of the fact that most of the millions of tons of polyurethane that is produced each year ends up in landfills because it is difficult to recycle and few facilities deal with it.

These enzyme discoveries are plastic polymer specific, but they open up the possibility that other microbial enzymes may exist that can break down some of the most pervasive plastics—polyethylene, polypropylene and PVC. Any plastic at all, really.

There has been a great deal of research in other fields as well, all searching for ways to break down plastics and recover reusable by-products. There is growing interest in technologies that use the energy stored in plastic's hydrocarbon structure to create fuel, known as chemical recycling. It offers a way of using the limitless plastic waste we currently have, creates economic opportunities and takes some pressure off the environment. As of 2020, the American Chemistry Council estimated that the US could sustain six hundred of the facilities required to support these "advanced" recycling technologies, generating close to thirty-nine thousand jobs and $9.9 billion in economic output.

Power Technology, a news and information hub for the global energy sector, reports several advancements in the plastic-to-fuels technologies. These include using polyethylene to create petroleum and other fuels, using HDPE to create plastic crude oil, using any plastic (and it doesn't have to be clean) as a source of hydrogen, and a process to turn plastic into a sulphur fuel—using the by-product to create a cleaner "ultra-low" sulphur diesel.

Critics of the plastics-to-fuel industry say that the use of these fuels may distract from the development of green energies and point out that they are just other forms of fossil fuel. Although I agree with this assessment, I would also argue that they could provide a stopgap measure while we transition away from petroleum-based energy toward green energy sources—thereby using the petroleum products that already exist, rather than exacerbating the climate crisis by producing more of them.

Recently, scientists have been experimenting with thermal processes to extract hydrogen energy from plastic—a cleaner form of energy, the by-product being water. There have been important advances in this direction, and a study published in late 2020 revealed a new technology that holds great promise. In a simple one-step process, it uses microwave energy and a catalyst to produce large volumes of hydrogen gas (H_2) and a residue that is primarily carbon nanotubes.

A catalyst is simply something that accelerates the rate of a reaction. In this experiment, they used a kitchen blender to pulverize the plastics from milk containers (HDPE), plastic bags (LDPE), food wraps (polypropylene) and plastic foam (polystyrene) into one-to-five-millimetre pieces. They mixed the plastic pieces in a 1:1 ratio of plastic:catalyst (of iron oxide and aluminum) and subjected the mixture to microwaves of the same power and frequency as your average household microwave oven. The mixture was "cooked" under 1000 W of microwave power for three to five minutes. This is just slightly less complicated than my muffin-in-a-mug recipe.

The method is genius in both its simplicity and its valuable outputs. Both the hydrogen gas and the carbon nanotubes—which can be used in electric batteries—are clean, green energy sources. This technology is still in the development stage but, if scaled up, could become an important part of the plastic solution.

What is even more exciting than these advances themselves are the possibilities they open up for a cleaner future.

———

These findings in no way mean we should take our foot off the gas. Governments and industry alike depend on our support for their political and financial success. This gives each person power to effect change. We need to speak up, lobby government to create policies and enact legislation that protect the environment and hold producers responsible. Demand that our tax dollars are spent on research and innovations that address the problem. And while we wait for the snail's pace of legislative change, we can lobby industry to clean up its act, to find and use alternatives to plastic. We can use our purchasing power to send the message that we don't accept plastic-coated products; we can choose to support the growing number of businesses that are avoiding single-use plastic or are coming up with innovative products that do not require plastic packaging at all. For those with an investment portfolio, make sure it reflects the values and ethics that will support the future you want to invest in.

We can support universities, research institutions and reputable environmental organizations that are looking at a range of solutions, from innovative ways to clean up plastics to replacing plastics with environmentally sustainable alternatives. And of course, we can clean up our own act at home: refuse single-use plastic; refuse any plastic where there are alternatives; buy clothes made of natural materials. Buy second-hand. Buy less.

Chapter 15

New Beginnings: Land of Ice and Hope

Antarctica is legendary. Mythical, even. Glaciers pierced through by jagged mountains peaks, calving glaciers the size of small countries, unforgiving wind and hull-crushing ice. Travel to the southernmost continent is an expensive undertaking. It can be the trip of a lifetime, the reward for years of saving. For seasoned travellers with deeper pockets, it can simply be an opportunity to experience something new. But most have a compelling reason—the seventh continent to tick off their list, a chance to see penguins, leopard seals, glaciers. To visit the places on the map that have fuelled the imagination—stories of whalers and explorers, driven by some potent cocktail of poverty, desperation, a sense of adventure, greed, ego, hubris. Places explorers had approached with unbridled anticipation, places where they barely managed to survive. Or from where they disappeared, never to be seen or heard from again. Impotent against forces of currents and ice. The bitter cold.

In 2017, before my epic back-to-back expeditions from Antarctica to Newfoundland, I had my first opportunity to travel to the Southern Ocean. Visiting Antarctica had been a lifelong dream, to experience first-hand the only uninhabited continent on earth, where marine wildlife thrived. It held some of the promise of the pristine wilderness I had imagined as a child. I had spent a lifetime with seabirds in the North Atlantic—my backyard. But the species that flourish in the Southern Ocean were almost entirely different. It was a rare gift: to be able to start all over again.

Though the place held a sense of wonder, I no longer possessed the naïveté of a child. I knew of the brush with extinction that whales and seals had endured in the past. And of the novel threats that marine plastic and climate change pose, not only to marine mammals but to seabirds found nowhere else in the world.

Whales and seals are now recovering—and some are even thriving—because of the interventions that came in time to save them. In 2017, Antarctica was not only a place of wonderment but also a site of inspiration and hope for the future.

———

It's a stretch to call yourself an expert on a group of species you've never seen before. But let's face it, the pool of seabird experts with experience in the Southern Ocean and Antarctica is pretty small, and I had decades of experience in the North Atlantic, which counts for a lot. Hurtigruten was looking for a seabird biologist to work on one of its small expedition cruise ships, available to travel to the Falkland Islands, South Georgia and the Antarctic Peninsula for six weeks over the 2017 Christmas season.

Most people with a decade or two of experience as naturalists and guides also have families at home, making it difficult if not impossible to be away over the Christmas season. But I cut a deal with my own family. We would celebrate Fake Christmas, 2017. The tree-decorating party, the fights over who would put up the lights, the unpacking of the nativity scene, the turkey dinner, the presents—all were geared toward December 4. Not one of them wanted to deny me this opportunity, and so it was decided.

There are over a hundred seabird species found in the Southern Ocean, and I was familiar with only a small handful, the long-distance migrants—those that spend part of their lives in both the North Atlantic and the Southern Ocean. Arctic tern, sooty shearwater, great shearwater. Long-tailed, parasitic and pomarine jaegers. That left a very long list of species I had never encountered. I spent the months leading up to the trip poring over field guides, reading and rereading the Antarctic bible, *The Complete Guide to Antarctic Wildlife* by Hadoram Shirihai. Watching videos, studying colour, size, behaviour, flight patterns. Poring over distribution maps, so I could chisel the list down to the species I was likely to see. I made a giant poster of all the seabirds of the Southern Ocean and hung it in my office, hoping for some passive transmission to my brain—the tail length of a slender-billed prion, the chunky head of a white-chinned petrel. Birds I had never heard of. By the time I left St. John's, I was as prepared as I could be.

On December 15, I boarded a plane headed to Ushuaia, Argentina—a small town on the southern tip of South America and the hub of expedition ships heading to the Antarctic Peninsula. After twenty hours of travel and a cramped overnight economy flight, I found a person holding a sign with my name and those of my fellow team travellers near the baggage area. Apparently on this Christmas trip there was to be a Holly, a Rudolph and a Claus. There were exhausted but cheerful exchanges among the arriving expedition team members—enthusiastic reunions between people who had worked and travelled together before, and a mutual sussing out of everyone else that had never met but would be living closely with for the next six weeks. The driver delivered us to our hotel with further instructions: a taxi would pick us up at eight o'clock the next morning to deliver us to the ship. Be ready!

The itinerary we were embarking on followed the path of the Scotia Arc, which is an extension of the Andes mountains that submerges below the ocean surface. If you were to stand on the southern tip of South America, face toward Africa and stick out your tongue, you could picture it: the mountain range following the outer edges of your tongue forms the Scotia Arc. Toward the end of your tongue, the mountain range rises out of the ocean, creating South Georgia and, at the very tip, rounding the curve, a string of islands called the South Sandwich Islands. The mountain range dips below the ocean surface once again until it rises to form the islands off the Antarctic Peninsula at the beginning of the continental shelf of Antarctica. The tongue's centre is a deep trough, gouged out by the force of the raging sea that flows west to east, uninterrupted, around the entire continent until it reaches the narrow bottleneck between South America and Antarctica. This narrowest part of the channel is the infamous Drake Passage, with the roughest seas on earth.

On disembarkation/embarkation days, the ship is a beehive of activity. Passengers leave, followed by some of the expedition team at the end of their contracts who are heading home. Before the new group of passengers arrive, there is a lot to do in a short time. As a first-timer with this company and in this place, I was on a sharp learning curve. Rudolph, a tall, lean Chilean who had been working as a naturalist with the company for many years, took me under his wing. He was relaxed and spoke slowly, his voice rising and falling like the lilt of an ancient ballad, always through a half smile. At the centre of all the frenetic activity, he left the impression that there was no rush—the perfect antidote to my intrinsic fluster, and I was grateful for it.

We boarded the ship, turned in our passports, got our room keys. News spread by word of mouth among the team, called over a shoulder in halls and stairways: there would be a meeting by the fireplace in an hour. There was an expedition leader and an assistant expedition leader, but they were too busy to present themselves to the newcomers before the meeting. The assumption being, if you are on the expedition team, you are independent and resourceful—figure it out. A frenetic sequence of events, typical of the first day of such trips, unfolds:

Jobs are assigned and everyone scatters. New arrivals pick up their uniforms, radios, first aid kits. Prepare the information boards—find relevant materials for the region, hang maps, print the forecast. Organize the jackets provided to the guests: remove them from their boxes and individually wrapped plastic bags, arrange by size, and set up traffic flow in and out of the room. Break down and get rid of all the packing boxes and plastic bags, destined for the ship's incinerator. If you are new, get oriented. Change into your uniform to greet the guests as they arrive, directing them to the hotel and administration desk, chit-chatting, answering questions, then directing them to the room where they can receive their complimentary expedition jacket. The jacket crew help guests find the correct size, offering advice on how much room should be left to accommodate layers, and excuses for why perhaps the jacket size they wanted didn't quite line up with their girth. Tact. Still others receive luggage on the cargo deck, working with the deck crew to sort by name and room number, further sorting into deck levels for delivery to the owner's room. A luggage detective attempts to track down bags that have not appeared, have lost their tags, have inadvertently been delivered to the wrong room or are hiding behind a door. Prepare the lounge for the mandatory fire and safety drill. Tick names off as people enter the room, directing them according to their assigned port or starboard position. Make sure all passengers are present and accounted for. Maybe you have to don the survival suit—check your list to see if that's one of your jobs. Guide the guests out of the lounge to their muster stations. Drill over. More chit-chat. Dinner and then the evening introduction of the captain, officers and department heads, and a quick parade of the expedition team members—everyone there to make this trip the experience of a lifetime.

Finally, a quiet moment. I made a surreptitious trip to the aft deck with my binoculars, alone. We had slipped the lines and were steaming slowly through the Beagle Channel, heading east toward the Atlantic. In the distance, the slow, graceful, meandering flight of a large seabird. I breathed in deeply, collecting my scattered attention for this singular moment. And I thought—*I'm screwed.*

I was embarking on a journey that would cover about six thousand kilometres of open ocean. I hadn't spent a lot of time preparing for the birds of the Beagle Channel. I had read a few lists, scanned some material, but given the scope of what I needed to learn, it seemed inconsequential. Yet here I was. The Beagle Channel (named after the HMS *Beagle*, the ship from which Charles Darwin famously explored the region) is a narrow ribbon of water that winds for

240 kilometres and joins the Pacific to the Atlantic. Not an insignificant distance to travel.

As the ship moved toward the centre of the channel and picked up speed, those graceful mysteries came closer and started following us. The pale face and ice-blue, reptilian eyes—unmistakably, a giant petrel. Pink bill tip: northern giant petrel. Phew.

Others joined the first, following the ship. During the last hours of daylight, I studied their flight pattern, size, colour variation. The clunkiness of the bill. How they approached the ship closely then peeled off, making large loops before returning to follow again. There are well over fifty species of tubenoses in the Southern Ocean—that large group of species that have an extra tube on the top of their bill, mentioned earlier for their keen sense of smell. The tubenoses dominate the Southern Ocean and range in size from the robin-sized prions and storm-petrels to the enigmatic master of the world's oceans—the leviathan of seabirds with a wingspan of 3.5 meters—the wandering albatross. I wanted to know the giant petrels well, so that I would recognize immediately if something different came into view, something that *wasn't* a giant petrel. At this point in my Southern Ocean birding experience, you were either a giant petrel or you weren't.

Darkness fell. The remaining mysteries and challenges of the journey through the Beagle Channel passed under the cover of darkness.

At dawn the following day we were in the wide-open ocean. The thunder of wind, muscular seas and heavy skies. Our first stop on the itinerary was the Falkland Islands, and the voyage to get there would take the better part of two days.

Sea days can be a challenge, especially if the seas are rough. At the beginning of any trip there is a restless anticipation for what is to come. Most people are in it for the landings, not the commute. There is a period of adjustment—of finding sea legs, battling motion sickness and the sleepiness that the motion brings. But I love the languor of sea days. Between the general tasks of necessity, I have the opportunity to spend hours outside on deck. On this first trip to an ocean of such renowned abundance, no one was more thrilled to be there than me.

I was on deck at first light, before official duties started, before "naturalists on deck" was featured on the daily program. The serious birders were already there, in quiet anticipation. Binoculars and cameras poised, waiting for the first silhouette to appear in the dim light—they rarely left the deck. They understood how special this remote part of the planet was and they were not going to waste a minute anywhere else. The opportunity to add dozens of new species to their life list, if they were patient and vigilant. These were the people I knew for certain I would never see at one of my presentations. How to Identify Seabirds at Sea? Please.

I have never been a dedicated lister. I have kept lists and lost them. I can't tell you how many species I've seen this year (year list) or in my life (life list). I can't tell you how many I've seen in Newfoundland or Antarctica. I have the information in notebooks—sometimes it's my job to know. And I like to look at my notes sometimes, but more as a trip down memory lane, or for the practical purpose of knowing who I might expect to encounter in preparation for an upcoming trip. For some people the list is more important than the birds. For me, it is the exact opposite. I have on occasion chased rare birds around with my birder friends, who I would say are passionate about both the birds and the list. But for me, it was more about the good company and the thrill of the sport, which I could easily engage in. Decades ago, a clay-coloured sparrow—a rarity in Atlantic Canada—was seen in a local suburban neighbourhood. I joined the search to find it, and when we did? Meh. A drab sparrow. I have nothing against drab sparrows, but as far as experiences go, give me a common loon on a lake. Or a drab sparrow sneaking around in the underbrush of a bog. I prefer to see birds thriving, in the places they are meant to be.

Dawn has its own temperament. It demands silence and solitude. Exchanges in hushed tones. Although it is my job to interact with people, it is just as important to know when to stay away. Early dawn on the first morning was as much about gauging the birders as the birds themselves. A polite good morning, and a position on deck that afforded good views and a generous personal space. The giant petrels had been around since the previous day and were not commented on. They had been joined by Cape petrels—a species that would become a faithful companion for most of the journey. The Spanish name is *petrel pintado*, meaning painted petrel. A much more fitting name for the white canvas, the head dipped in black, an intricate pattern of black sweeps and splotches across the back, the wings, the tail. A beauty.

Before long, the silence broken. "Black-browed albatross at seven o'clock"—a voice, raised just above the wind, indicating the position of the bird, the bow being twelve o'clock. With the wingspan that exceeds the height of a tall man—two and a half metres, held rigid— the albatross approached low over the water toward the ship, like an airplane approach on a runway. Then, an undetectable change of angle in the wings, sending the bird in a lazy arc skyward, before turning to descend once again, accelerating toward the ocean surface. The smooth, effortless S-curve drawn repeatedly, just above the ocean surface: dynamic soaring. Albatross are known for it. They can cover hundreds of kilometres without taking a break, using the energy of the wind and gravity instead of their own. The wings locked in position, a mere shift in the angle—the perfect design.

I was glad for the solitude in that early morning light. My first albatross, on the open Southern Ocean. Exactly where it belongs.

What is known as topography on land—contours that describe the shape of the landform's features—becomes known as bathymetry under water. The Falkland Islands are part of the southeastern portion of the South American continental shelf, known as the Patagonia Shelf. They are a group of over seven hundred islands that rise out of the water about five hundred kilometres from the closest point in South America. The surface topography is not dramatic, mostly low, rolling, windswept hills. The dramatic forces are happening below the surface.

The edge of the Patagonia Shelf is called the Patagonia Shelf Break, and this is where the waters of the continental shelf meet the cold, nutrient-rich sub-polar waters of the Malvinas Current (also known as the Falkland Island Current) that originates in Antarctica. This zone of mixing is called the Patagonia Shelf Break Front and is one of the richest and most productive marine areas in the world. The Falkland Islands are about 250 kilometres from the main shelf break front, but there is a zone located along the 200-metre contour line that marks the steepest surface temperature gradient that characterizes the front. This 200-metre-deep intrusion wraps around the islands like a scarf, and takes the rich Malvinas Current with it. The marine life it supports is breathtaking.

The Falkland Islands are situated at the same latitude in the south as Newfoundland is in the north. Islands, surrounded by a cold, rich ocean current—a scenario I am very familiar with, though the marine life they support is very different.

Our first stop was West Point Island, home to a colony of black-browed albatross, southern rockhopper penguins and elegant imperial cormorants. The expedition team landed on the island ahead of the guests, as is standard for all wildlife landings, to set up cones and flags indicating the path to be followed, and mark points guests should not travel beyond—for their safety, to protect wildlife and artifacts, or both. The highlight of this stop was the black-browed albatross and southern rockhopper penguin colony. I hurried ahead with the assistant expedition leader, Rike, and Delphin, a French naturalist, to get oriented.

From a distance we could see the colony against the backdrop of an intensely blue sky, the occasional scattered clouds racing with the wind. At the edge of the tussock, the trampled muddy ground, covered in a remarkably symmetrical arrangement of nests, sloped steeply toward the cliff edge.

The trail leading to the edge of the colony weaved through dense mounds of tussock grass, growing well above our heads. We stepped into the tussock and lost sight of the birds. Rike wanted us to get oriented in the confusing maze and to be aware of the particularly sensitive places.

"The penguins travel under the tussock grass—watch your step," Rike warned.

"And make sure the guests know this as well."

Although the tussock grass provides a visual barrier to the colony, there are places along the established trail where small gaps between the mounds are protected by only a thin veil of the tall grass. I crouched low to the ground and very slowly reached between the blades, parting them with my fingertips. I found myself eye to eye with an intimate and timeless ritual that was unfolding, completely oblivious to my gaze.

A pair of albatross sat on a tall pillar of mud built up from the ground, their nest. The pair were in the midst of a raucous greeting—loud, high-pitched, rattling calls, like a plastic kazoo, tapping their heavy bills together, then bowing their heads low. Each time a partner returns from a foraging trip at sea, this ritual occurs. If not exactly romantic, it seals the bond between the pair.

This noisy, animated, social creature was entirely different from the loner I had watched for hours at sea. The dark feathers of the brow, which looked slightly sinister from a distance, looked more like the intricate work of a make-up artist at close range—the blending of greys and black, creating a smokey eye. The bill, yellow at the base, deepening to an intense orange near the tip, like a sunset. At sea, albatross appeared weightless; on the ground, they looked every ounce of their nine pounds.

Once the greeting was completed, the pair changed roles. The parent that had been on the nest and caring for the chick for the last couple of days was now free to go—lifting its wings awkwardly in the relatively confined space of the nest and pointing into the powerful wind for takeoff. Once in the air, it hangs almost playfully, riding the updraft from the sheer cliff below, feet dangling, before turning, feet tucked, and heading out to sea on a solo journey of hundreds of kilometres in search of food for itself and its young chick.

Back at the nest, a slight jostling and rearranging of weight, to accommodate a small white downy head, a feisty and impossibly fragile chick, imposing its will, jabbing its parent's bill—demanding to be fed. On cue, the parent's gape now wide open, the chick sticks its head down the parent's throat and receives a slippery pink meal of squid.

The same scene plays out thousands of times over in identical nests throughout the colony spread along the windy cliff edge, the racket and stench carried out to sea.

Between the mud pillars, the southern rockhopper penguins. A striking and comical bird, the bright-red eye and matching beak; the spiked yellow-and-black feathers adorning the head, blowing this way and that in the constant wind. Like all penguins, they stand upright, accentuating their round belly. They look like aging rock stars—well past their prime but unwilling to give up the stage hair.

Making their way between the mud and stones, they hop, of course. From the sea all the way up to their clifftop colony. Their nests are a less showy arrangement of mud and stone, interwoven with tussock grass. The rockhoppers and albatross make perfect neighbours—which is to say, they completely ignore each other. Rockhoppers breed in large colonies, and although they do not require the presence of albatross, they are a definite asset when aerial predators are around. At this colony, a striated caracara, a predatory, hawk-like bird opportunistically in search of exposed eggs and chicks, was patrolling. But no caracara would mess with an albatross, a fact that is not lost on the rockhoppers.

The black-browed albatross and the rockhoppers have taken completely different approaches when it comes to reliance on the ocean for food and survival. Like all seabirds, they spend the majority of their lives at sea, coming to land only to breed. But how they have evolved to spend that time is completely different.

The grace and efficiency of albatross flight enables them to fly hundreds of kilometres, scanning the ocean surface over vast areas for fish, squid, krill and other prey that venture close to the surface. They will also take carrion on the surface, and follow long-liners, stealing squid from the hooks as they rise to the surface.

For rockhoppers, the story is very different. Evolution flattened the long bones of the wing, fused and shortened them, into what functions more as paddle than wing. Penguins gave up the ability to fly in exchange for becoming champion swimmers and divers, manoeuvring through the water with speed and agility. Like the black-browed albatross, rockhoppers eat fish and squid, along with krill and other prey. But they are diving for theirs—up to forty-five metres below the surface, out of reach for the albatross. In this way, they avoid competition with each other. The best of neighbours.

"Holly, it's time to go . . ." A gentle nudge from Rike. "The guests will be here soon."

The next stop was Stanley, the capital of the Falkland Islands. It was typical of most small working port cities, where fishing is central to the local economy. Along with fishing boats, there were bits of discarded fishing line and rope, and small sheens of oil. Wind-thrown bits of garbage. Cigarette butts. Gulls. None of this was unusual, except that I had a heightened perception of how important the area was for marine life. Having experienced so much of it for the first time upon approach to the islands made me acutely aware of their significance.

The Falkland Islands have not only the largest population of black-browed albatross in the world, but also the largest gentoo penguin population. The islands are home to a whopping twenty-two seabird species, including three species of penguins, three albatross and four petrels whose populations are listed by the International

Union for Conservation of Nature as species of conservation concern, ranging from near-threatened to endangered. All the listed species exist only in the Southern Ocean. The intimate encounters with albatross and rockhoppers at West Point the day before made it personal. The trash in the harbour suggested a casual disregard for the magnitude of this area's significance, and it made me uneasy. Of course, for the residents, they were living their lives in the way that people do everywhere—no better, no worse.

From the Falkland Islands, the ship continued southeast toward South Georgia, another two and a half days at sea. The temperature was dropping as we approached the edge of the Antarctic Circumpolar Current. Here the warmer Atlantic Ocean wages battle against the frigid Antarctic Ocean, creating the Antarctic Convergence—that extraordinary region where marine life fluourishes. The Antarctic Convergence encircles the Antarctic Ocean, defining its boundary against the Atlantic, Pacific and Indian Oceans. South Georgia was 1,800 kilometres from the nearest landmass—and we were sailing toward it.

There are twenty-five seabird species that breed in South Georgia. Six species of penguins and four albatross, including the wandering albatross. Fourteen species of petrels and prions, the brown skua, the Antarctic tern, the imperial shag. South Georgia is home to the largest populations in the world of light-mantled albatross, grey-headed albatross, northern giant petrels and white-chinned petrels. An exotic list that most people have never heard of, and only a lucky few have ever seen. The region is astounding not just for the list of species it supports but also the population size: there are over 30 million pairs of seabirds breeding on the island. All but the imperial shag are pelagic seabirds, meaning they are completely at home in the open ocean—soaring above it, resting on its surface or diving below for food.

The pelagic seabird enthusiasts on deck were not going to miss a moment of it. After long months spent dreaming of and preparing for this opportunity, it had arrived. Small groups huddled against the battering winds, hands warmed against a cup of coffee, taking occasional brief breaks to thaw frozen fingers. There is a generosity among birders, in their pooling of observations and experiences. As nuances of patterns and features were noticed on these novel species, they were shared and discussed. New birds were pointed out by whoever saw them first. A fellowship of the like-minded.

The list mounted—grey-headed albatross, South Georgia diving petrel, more prions—each new species a collective celebration. Gentoo penguins, porpoising through the water. I did a double take the first time I saw it. I'd never witnessed a bird porpoise before, and at first I thought it was a fur seal. They travel this way for the same reason dolphins and porpoises do—to continue breathing while on the move, using the forward momentum to propel themselves out of the water and take a quick breath before submerging again. An efficient system for species that breathe air but live in the water.

Far off in the distance, finally. A stark-white, barrel-chested giant, seeming to defy gravity, its wings a pencil line against the sky, disappearing from view over the ocean before appearing again above the horizon. That distinctive pattern of dynamic soaring drawn over the ocean surface. At ten knots, the ship was being outpaced. It drew closer, before finally overtaking the ship and slowly disappearing into the distance ahead.

A wandering albatross.

In that moment there was no sound. No wind, no voices, no engine. No smell of ocean spray or ship's fuel. No bracing cold, no frozen fingers. Just the arc of its flight. The slow, confident steadiness of it. The white wings, with patches of black bleeding through and deepening toward the trailing edge. The enormous head. The swooping curve of its bulky bill, a pale pink. The exposed underside as it arced toward the sky; pure white, with just the finest line of black on the trailing edge of the wing; the very tips dipped in ink. It looked too heavy to fly and yet it was overtaking the ship with an unhurried ease. Capable of sustaining flight speeds up to 130 kilometres per hour for hours on end, circling the globe in forty-six days. The three-and-a-half-metre wingspan, over twice my height.

The American ornithologist Robert Cushman Murphy famously describes his personal encounter while aboard the whaling ship *Daisy* during a hunting foray to the Southern Ocean in 1912: "I now belong to a higher cult of mortals, for I have seen the albatross."

I don't know about mortal cults, but I certainly felt grateful. For the life choices that I'd made, the circumstances that brought me here. To my family, for faking Christmas. I am never going to be wealthy or have a dental plan. Those are the choices I've made. For experiences like this—almost two thousand kilometres from the nearest continent, on the Southern Ocean where conditions are so inhospitable by human measure yet exactly, unimaginably, perfect for a multitude of species, including the wandering albatross—the most revered of them all.

I have good teeth, anyway.

On our final approach to South Georgia we encountered the Shag Rocks, a group of small, unvegetated islands at the western tip. This area was much anticipated and greatly hyped for the sea life it attracts. If you look at a map of the ocean currents around South Georgia, they conjure the serpents of Medusa's head, a frenzy of currents whipping and curling out in all directions. Aside from the imperial shag colony for which the islands were named, thousands of seabirds, seals and whales congregate to feed in the rich waters fed by the violent upwellings the currents create. If you are lucky, a mix of the large whales can be seen by the hundreds here. If you are not lucky, you will be treated to a fog bank, a feature that is frustratingly common in areas where warm and cold ocean currents collide. It was not our day. The fog shrouded the rocky islands. We waited for an hour or two, giving the fog a chance to lift, but it didn't budge. Many were disappointed. I have found myself in fog banks many times and know that disappointment. But I also felt a sort of reverent admiration. No matter how far you go, or how hard you try, sometimes the sea just doesn't give up its secrets. The power of nature over human will.

Besides, the most highly anticipated South Georgia event for me was yet to come: a visit to Fortuna Bay, a colony of about three thousand breeding pairs of king penguins—the stuff of my childhood fantasies. King penguins are second only to Antarctica's emperor penguin in size, standing at 85 to 95 centimetres (about three feet)— tall enough to lay its head on your dining room table. It is the most brilliantly coloured of all the penguins, with intricate patterns of cayenne-orange earmuffs and a matching flourish along most of the lower bill. The black head that gives way to an intense smear of burnt orange under the chin that melts to a faded yellow, then white. As though the head is rusting, the stain running down its throat. Unlike most penguins, the back of the body is not black but a deep to silvery grey, depending on how the light catches. A magnificent and regal-looking penguin, deserving of its name.

The expedition team boarded the first tender boat to scout the coast-line for a place to land. The beach was covered in fur seals—mothers with clusters of young nearby, large males spread throughout the rookery, each defending its territory. The bull fur seals, at 188 kilograms, are significantly larger than the 45-kilogram females. Interspersed amongst the fur seals were the behemoth elephant seals. The females, up to four metres long and 800 kilograms, were dainty in comparison with the six-metre-long, 3,700-kilogram males. In various shades of grey and brown, they looked more like geological features. Luckily, their pace of movement was about the same. They were merely resting and moulting and were not much interested in or disturbed by us.

Tessa scanned the beach, looking for an opening in the fur seals where we could land safely and avoid disturbing the rookery as much as possible. The colony had expanded since the last time she'd been here; it was a challenge. I stared in amazement. We are going to land here? How? Where?

A gap was found and the driver landed the tender boat on the beach. We were quickly schooled in how to ward off any aggressive animals.

"If you are being charged by a fur seal, place your walking stick on the ground in front of them. They won't proceed any further," Tessa said. "Don't touch them, of course," she added.

I felt a little dubious about the whole thing. Could that possibly work? What brave fool was the first to think that over four hundred pounds of hormone-addled and surprisingly nimble fur and fangs could be thwarted by a collapsible hiking stick? And was I going to be the fool who trusted it? Yes, I was.

The team assembled on the beach and were issued instructions to find and flag the path of least disturbance through the rookery, leading to the king penguin colony. A few penguins were sauntering along the beach, paying no heed to us or the seals. Two thoughts:

"Wow!" and "Focus." We were to lead small groups of passengers along the path, "protecting them" from any potential fur seal attacks along the way. Tessa briefed the passengers as they landed. I gathered my group and headed out.

"Stay behind me," I reminded them, "and we'll see if this really works."

I didn't say the second part out loud. Before long, a male broke away from the rookery and came bounding at us, growling. I placed my walking stick on the ground in front of me. He stopped. I kept the stick on the ground and ushered my group to proceed behind me, away from the assailant. I'll be damned. It worked.

We continued along the beach, dodging fur seals, until we reached the edge of the king penguin colony. What a spectacle— thousands spread out through the valley, with a backdrop of glacier-capped mountains. From a distance, the colony looked like a complicated knitting pattern of greys, whites, browns and orange, the downy chicks a popcorn stitch.

Depending on species and size, seabirds vary in the length of time it takes to breed and raise their young. For most species, chicks are raised to independence within a single year. Colonies synchronize their chick rearing over a period of months in the summer, abandoning the colony en masse after the breeding season. King penguins are the exception. They take a full fourteen to sixteen months to complete the breeding cycle, so that a pair breed only two out of every three years. What this results in is a population of mixed ages and stages that occupy the colony year-round. The presence of the relatively warmer currents from the North Atlantic and the interaction of the currents around South Georgia mean that it has open water and an adequate food supply all year long.

In the centre of the colony there were adult birds with young chicks covered in a fluffy light-brown down, their high-pitched, barely-there calls carried on the wind. Toward the edges, an array of adults and more independent chicks, gathering in groups called crèches, the French word for nursery. The crèches were attended by a few adults, taking turns. Standing almost as tall as their parents, some chicks were still covered in the long fluffy down, resembling giant plush toys. Slightly older chicks were moulting, large hunks of down missing, exposing patches of the sleek plumage of their parents. The awkward teenage phase. None of them were shy. The rule is, you cannot approach penguins any closer than five metres—but there is no rule against them approaching you. Playing with the straps of your backpack. Giving your leg a curious poke. All the while, giving that baby bird call—at close range, a high-pitched, incessant chatter, like an excited child with a story to tell. The adults, though not shy, remained a little more aloof. Turning their backs; a pair trumpeting to each other. They had other things on their mind.

We left from Fortuna Bay to hike the last six kilometres over the mountains and glaciers of South Georgia to the whaling station at Stromness. These were the last six kilometres of Antarctic explorer Ernest Shackleton's brutal odyssey in 1914. After his ship, *Endurance*, was crushed in the ice, Shackleton and his crew survived six months on the ice floes before making landfall on a tiny spit on Elephant Island. With little hope of being rescued there, Shackleton took the largest lifeboat and five of his crew, sailing 1,300 kilometres across the open Southern Ocean before sighting the southern shore of South Georgia. Sixteen days in an open boat. And still a mountain range between the men and safety. Shackleton and a crew member walked from the southern shore with only a length of rope and shoes with nails driven through the soles to grip the glacier ice. Against all odds, they made it.

For us, it was a steep climb through a mountain pass, but without the worry of survival we were able to appreciate the stunning, rugged peaks and glaciers of the surrounding landscape. At last, we reached the much-anticipated summit that offered the first glimpse of Stromness, down in the valley below. The rusted-out graveyard of dilapidated buildings that remain. The abandoned infrastructure built to support the processing and rendering of oil from whales slaughtered in the bountiful waters around South Georgia and Antarctica. In our minds, we were all transforming the image below to the scene that we imagined unfolding before Shackleton and his men, a bustling beehive of activity—hacking, hauling, burning, sweating, smoking, cursing, laughing, below a constant curl of black smoke, lifting. Probably a whale being flensed on the beach, men working with sharp blades to peel the skin like a ripe banana. The stench of blood and smoke. The racket of noisy industry—music to their ears. In this place of bloody and brutal work, they had finally reached a safe haven.

We were moved by what we saw and what we imagined. This place, where Shackleton's survival was ensured. This place, born to support a whaling industry that saw the slaughter of hundreds of thousands of whales before it was all said and done. This place of salvation and slaughter. And a testament, too, to the mess we leave behind.

We started the steep descent along a stream that led down to Stromness and were met by gentoo penguins trudging in the opposite direction, toward their nests in the higher reaches of the valley. A cool grey mist hung in the air, a gloominess that felt completely appropriate. The buildings in Stromness are now off-limits; between asbestos and falling detritus from the deteriorating structures, they've been deemed unsafe for human occupation. The fur seals have reclaimed the space, the rotting buildings providing them with cover and protection from the wind. They are now simply features of the landscape to be explored and used to advantage, with the added bonus of no human interlopers. But even here, there was no escaping the reach of plastic: one fur seal on the beach with a section of gillnet around its neck.

I was assigned to an area at the base of the meadow, a safe distance from the back of the station's infrastructure. I wandered to the edge of another stream that formed the farthest boundary, for a moment of quiet reflection. In it, several young elephant seals, or *beaters*, were bathing. Enormous adolescents blowing bubbles, rolling, moaning, burping and farting in the mud—a much-needed moment of comic relief.

It was another two days at sea before we would reach the continent of Antarctica. A few seabird species remained faithful companions of the ship—Cape petrels, southern fulmars, giant petrels. Others came and went. Late on the first day at sea, a final parting gift from South Georgia: a pair of light-mantled albatross, flying in synchrony, like pairs figure skaters. Dipping and turning in perfect time with each other, a rhythm that is bred in the bone. Most seabirds are some version of black and white, but not these. If you took the coat of a Siamese cat and draped it over the body, you would have the plumage of the light-mantled albatross, the deep browns melting to a subtle pale shade at the nape of the neck. They were stunning in their subtlety. A semicircle of white around the eye gives them a permanent look of mild astonishment, as if they are surprised to see you. Touché. They breed on South Georgia, but we had not seen them while we were there. They did not follow the ship, more interested in each other than in us. Something about the encounter felt mystical to me. I know very little about the species, by choice. I have no desire to dissolve the magic with metrics. Some things should remain a mystery.

The water and air temperature grew steadily colder as the ship headed south toward Antarctica. Most of the seabirds we left behind to the rich waters of the Antarctic Convergence and the diverse abundance of prey it supports.

———

Antarctica is unlike any other place on the planet, in many ways. Including this one: it is not owned or controlled by any country but is governed instead by the Antarctic Treaty and associated agreements, collectively known as the Antarctic Treaty System. The original agreement was signed by twelve nations in 1959, but today there are fifty-four signatory countries. All with the commitment for peace and protection of the vast remote continent. Military installations are not permitted. Scientific co-operation and collaboration between nations is encouraged. It has been a stunning success.

Antarctica is a refuge for hope. It demonstrates not only that we have the potential to cause devastating destruction, but that we also have the capacity to work together, to rebuild and restore our damaged planet. That recovery is possible.

The Antarctic marine ecosystem is also unique in the world. It has less diversity than that of South Georgia, but it more than makes up for that with abundance. As in the Arctic, the ice forms a substrate on which ice algae grows. The ice sheet that forms around Antarctica doubles the size of the continent in winter, and this allows for a massive field of algae to grow under the ice, which is grazed on by a species of krill called *Euphausia superba*. This fatty, nutrient-rich little crustacean packs a punch. The estimated biomass in the Southern Ocean is over 400 million tons—about the same biomass as all humans on earth put together. It is the keystone species of the marine Antarctic food web. Whales, seals and penguins spend their summers gorging on them. Penguins and seals depend on the ready abundance to feed themselves and their young.

In spring, whales—blue, fin, sei, humpback, southern right, minke, sperm, orca—travel en masse from equatorial waters to the Antarctic to fatten up after a winter of fasting. Blue whales eat only krill, up to six tons a day. Roughly 16 million penguins and their young depend on krill as a primary food source in Antarctica, as well as around 20 million seals, give or take. By krill standards, *Euphausia superba* is humongous. Even so, it maxes out at sixty-five millimetres. It takes a lot of krill to support all this life. Which means a lot of algae. Which means a lot of ice. Warming oceans could threaten this balance, but so far, so good.

Our first stop in Antarctica was Elephant Island—Shackleton's rescue mission route in reverse. We were thrilled by the first sight of land through the grey mist—the mountains, the glaciers—though probably not as thrilled as the crew of the *Endurance*. The ship steamed slowly past Point Wild, the headland named after crew member Frank Wild. It marked the small bay and the tiny beach where the crew survived an Antarctic winter, waiting. At right angles to the beach, the glacier that covered the island spilled into the ocean. Any calving event from the glacier would have propagated large waves, flooding their beach and washing them out to sea. Frank Wild was the man left in charge on the beach. He was responsible for keeping the men alive and maintaining morale, a job at least as challenging as Shackleton's expedition in the small open boat. They ate whatever came close enough to kill. They were lucky to share the beach with a small colony of chinstrap penguins—easy prey, their frozen corpses stacked like cords of wood.

The expedition team attempted a landing on the beach, but the swell and battering chunks of sea ice made it too dangerous. Instead, boat cruises were organized, bringing guests as close to the beach as possible, navigating through the narrow channel that separated it from Point Wild to get to the other side. We cruised the length and breadth of the beach, and it didn't take long, cruel in its inadequate dimensions. The chinstrap penguin colony is still thriving there. A bust of Captain Luis Alberto Pardo stands in the centre, in honour of the captain of the Chilean navy rescue ship *Yelcho*. It looks like the centre of penguin town square, the chinstraps socializing around its base. They no longer had to worry about the appetite of a desperate crew. But there was another, much stealthier predator with an appetite for penguins lurking in the shallow, turbulent waters: a leopard seal was patrolling the beach, its nostrils periscoping above the surface for a breath before slipping below again. Barely detectible. Patiently waiting for a penguin to make a mistake.

"Keep your hands inside the boat," our tender boat driver warned. It was good advice.

There are several written accounts of leopard seals hunting the stranded men.

Penguins came and went from the water in small groups, unaware. Then there was an explosion from the water, like an erupting volcano—a leopard seal with a penguin clamped by the head in its jaws—a few violent shakes and the neck snapped, a fine mist of blood sprayed in an arc, small bits of flesh and fat like bullets scattered. Within seconds the skin was peeled off and the penguin turned completely inside out.

Welcome to Antarctica.

We remained close to shore for the next several days, weaving through the islands around the Antarctic Peninsula. On a clear day, the light is otherworldly—the intensity of it, reflected off every surface, covered in blinding ice and snow. The mountains, standing ridged and resolute, indifferent to the powerful katabatic winds, the constant thrust of ice, their dark faces a striking contrast to the ice, the ocean and the sky—each some combination of white and blue. Terrestrial vegetation barely exists. It is limited to 1 percent of the continent and consists almost exclusively of mosses, lichens and liverworts. The tallest plant in Antarctica soars to five centimetres. Plants are not a standout. If anything, it serves as a reminder of how foolhardy the notion, attempting to survive an Antarctic winter. It is easy to imagine that no one has been here before, the wind and snow wiping the slate clean.

Antarctic penguin colonies. After a lifetime of magazine spreads and documentaries, I was seeing them with my own eyes for the first time. Everything I had imagined, right there. A colony of gentoo penguins, their white ear patches like earmuffs, the bright-orange bill and more subtle pale feet. At the tip of each webbed toe, a surprisingly sharp nail, good for gripping ice. In the bay, leaping off islands of floating ice into the ocean, porpoising with ease through the water. A transition from swimming to standing in one agile motion, onto the beach. Ocean to land—from streamlined and slick to comically awkward—a bird designed for efficiency in water, sacrificing grace on land. They gather in groups, depending on each other for safety, and follow the well-worn trails they have carved through the snow—the "penguin highways"—exiting at the turnoff that leads to their nest, an elevated pile of stones. That endearing waddle. The group mentality at play, decisions imperceptibly made. *Let's go here. No? We're turning around? We don't like this? Ok— everybody turn!* Responding to some perceived problem up ahead, or maybe just a change of heart. The most moving thing for me is the penguins' complete indifference to us. At most, a brief assessment before carrying on, maybe diverting their course, in the same way they would detour around a Weddell seal basking on the beach. We were not a threat, but a novelty to be evaluated and dismissed.

Every day we made at least one landing. In the polar regions we haul a lot of safety gear ashore, a precaution against getting stranded due to weather or ice: barrels and bags, passed between the expedition team in a line: tents, blankets, food, water, first aid equipment, even a stretcher. Walking sticks, for guests, poles and flags for marking assigned trails and to identify "no-go" areas. Once the landing was prepared and the site laid out, each person in the expedition team took up the position assigned to them.

I was always at the penguin colonies. Once we were in position, there was usually a half-hour or more before the guests came ashore. This time, a gift of solitude. I spent it sitting quietly, moving slowly, camera poised. Ready for whatever penguin antics might ensue. A penguin carrying a stone in its bill, trudging hundreds of metres from the beach sometimes, the stone a gift to its partner, or added to the nest. Their nests are made of hundreds of these small stones, which act as weeping tile, allowing guano and water to run through, protecting both eggs and chicks from getting too wet. It is a lot of work to lug the stones, each requiring a separate trip to the water's edge. Unless you steal from your neighbour. Theft is a common sight, but no less amusing for it. It can be stealthy—literally behind the back of the sleeping dupe, completely unaware. Or taking advantage of the presence of an aerial predator nearby, the victim of the caper aware of the theft but deciding the danger of moving to defend its property would pose too great a risk to the defenceless eggs or chicks; the Antarctic skua tuned in to the action, anticipating a false move and a meal for its own young. A downy chick, wrestling for space and finally appearing from under its parent's breast—a wing stretch, a breath of fresh air, or jabbing at its parent's bill, begging for a meal. A throaty hee-haw exclaimed by one that crescendoes through the colony and fades into quiet again. The regular pulse of a penguin colony. Occasionally, a determined penguin making its way up the slope would stop just short of me, as though I had interrupted its train of thought. A pause before carrying on. If I was near a highway, I moved. If not, I let the bird decide its own route and remained motionless.

Conditions from one site to the next can be dramatically different. Neko Harbour is a sheltered harbour surrounded by high mountain peaks and a dramatic glacier face that frequently calves. The seas are usually calm, icebergs and pan ice moving gently with the currents. Humpbacks, minkes, Weddell seals. The comings and goings of the gentoos that breed there. In stark contrast, there is the chinstrap colony at Orne Harbour, a much more exposed location. The chinstrap penguins here nest high on the top of a steep face. A hard slog for penguins and people alike, but worth it for both. Because the top is windswept, it is cleared of snow and ready for nesting earlier in the season than protected areas.

The wind is not an issue for the hardy birds; in a blizzard they simply face away, snow piling up on their hunched backs. Skirmishes can be a little more dramatic on the steep slope, though. Once, an angry ball of adversaries rolled downhill, picking up speed, until they came up solid against my back. I broke their fall and perhaps gave them pause to reconsider. Shaking themselves off, they carried on their separate ways. Below, a leopard seal stalked the beach for penguins, with regular success. It took at least five that afternoon. Tiny Wilson's storm-petrels flitted just out of reach, deftly plucking flying globs of oil and small bits of flesh.

Not all of our Antarctic landings were at penguin colonies. On our third day we stopped at the old whaling station at Whalers Bay on Deception Island. Deception Island looks like any other island in the South Shetland archipelago, but it is not. A narrow channel on the southeast side—named Neptune's Bellows by nineteenth-century sealers—leads to the centre of the island, which is actually the flooded caldera of an active volcano. A Norwegian whaling station was established there to process whales in the relative protection of the mountains that rim the centre of the caldera. It operated there from 1912 to 1931.

Deception Island became the gateway to Antarctica and it has a colourful history. In 1928, Sir Hubert Wilkins oversaw the construction of an airstrip at Whalers Bay and became the first person to fly a plane in Antarctica. The Royal Navy established a base at Whalers Bay, to conduct scientific research and to reinforce territorial claims. In 1960, an aircraft hangar was built at Whalers Bay, and it became the centre for British flight operations in Antarctica. The British Base (Base B) remained active until 1967, when the volcano erupted, spraying plumes of black smoke and ash near a

Chilean base on the island at Pendulum Cove. The island was evacuated thanks to the Herculean bravery of helicopter pilots who flew in to the bases and airlifted people to the safety of a waiting Chilean ship on the outside of the island. No ship would sail into the caldera of an erupting volcano.

Whalers Bay was occupied again the next summer, but the visit was brief. Against all odds, Deception Island erupted once again. This time the eruption was much larger, and it destroyed the Chilean base and most of the buildings at Whalers Bay. The graveyard with its thirty-five graves. A monument honouring ten men. Everything was buried in ash and washed out to sea. Deception Island refused to be domesticated.

Wildlife had returnedfur seals and penguins basking on the beach, Antarctic terns breeding behind the old ruins. Dead krill washed up on the beach, evidence that the volcano was active, hot steam rising from underwater vents killing any krill that ventured too close. Along the shore, Antarctic terns scanned the shallow water, dropping to gather the easy pickings. Joining them, a scattered Arctic tern—their identical twin that breeds at the opposite pole, flying a round trip of about eighty thousand kilometres to partake in the Antarctic summertime abundance. To prove the existence of the internal heat, the guests were encouraged to bury their hands in the fine gravel. It is warm below the surface layer. As the tide fell, steam rose from the beach. The ecosystem had regained a natural balance, where life and death are hand in glove.

I scanned the beach, looking for something interesting for the guests to see close up.

A gentoo penguin carcass was splayed on the shoreline, pecked open by gulls. For a bird biologist, this is a guiding gold mine. I could show the short, densely packed feathers, unique to penguins, that offer superior insulation and waterproofing. The "wings" whose bones have flattened and fused for use as paddles while diving. The deep keel of the breastbone for large muscle attachment and the unusual extension of the rib cage, the rigid structure protecting the organs from compression during deep dives. All adaptations for a life that depends on cold-ocean creatures for its own survival. Shape-shifting to perfection. The whole Antarctic ecosystem is connected, and with a few props, the web can be woven.

But the place also felt haunted—the rust-orange graveyard of the old building set against the black ash that coats the glacier and the beach; the rising steam a reminder of the thermal energy lurking just below. A high-pitched whistle came from the Cathedral Crags rock-face near Neptune's Bellows. Cape petrels flew toward the top of the rock pillar and disappeared. A sound that was barely there, the birds disappearing into its face—it felt like a hallucination. Once the gateway to Antarctic exploitation, where the natural balance had been knocked perilously off-kilter, it's as if the island was sending a message: don't overstay your welcome. It instilled the same sense of wonder and dread I had felt so many years ago in the heart of the Torngat Mountains. Like Torngat's Lair, Whalers Bay is spectacular and strange and eerie. I was keen to be there, to see the place. And I was just as happy to leave.

———

During a quiet moment on deck on our last leg of the journey, one of the guests came over and stood next to me in a way that had come to feel familiar—a smile exchanged, both of us looking out to sea.

After a few minutes she said: "This is incredible. But it's not worth it. I'm sorry. We shouldn't be here," she said.

I knew what she meant; it is an issue that I also struggle with. I brought her comment to Delphin, the French naturalist I'd worked beside during the expedition. It was a question that he too has asked himself: if we care about the conservation of these places, maybe the best thing we can do is stay away. This thought doing battle with the concept that experiencing the beauty and majesty of a place makes you care more about protecting it.

I have not been completely sold on the latter, which is all too convenient for someone who makes a living from visiting these places. Still, if there is some truth to it (and I think there is), I wanted to bring as much awareness and conservation value to the experience as I possibly could. Getting people excited about seabirds is an important step, but I wanted to do more—a deep dive into the issues that threaten the survival of everything we had been so deeply moved by over the last several days.

"I think I'm going to prepare a presentation on marine plastic."

"Please do," Delphin said.

And the seed of this book was planted.

Since I began working on the book you are holding or listening to, I've heard a lot of stories, unbidden, from people who spend their lives on the ocean. Once they learn that I am researching marine plastic, it is as though they are speaking to a priest at confession, compelled to share the awful things they have seen and, sometimes, done. They want me to bear witness and maybe even give absolution (*you didn't know, you didn't mean to*): the plastic band from the end of a rubber glove wrapped around a turr's (murre's) neck, the bird alive but slowly choking to death; another turr with wings entangled in a six-pack ring, wearing it like a vest, immobilized and emaciated. A turr hunter described a bird he'd shot, some fishing line hanging out of the end of its bill that turned out to be the tail end of a ball of line compacted in its gut—he put his hands out to illustrate the size when he removed it—"bigger than the turr itself."

A fisherman who worked on a clam dragger for over thirty years told me that back in the day, they routinely dumped old gear overboard. On the draggers, it was mostly metal cables. "The metal just rusted and went back to the earth—but not that old rope." There was the admission from Vince, the lighthouse keeper at Cape St. Mary's, that he used to turf his garbage bags over the bill of the cape, not far from the gannet colony. Shaking his head with incredulity at the behaviour of his younger self. But he can be forgiven. And so can the dragger fisherman, and the guy with the six-pack of beer. And me too, with my pile of plastic that lined the counter in Sunnyside. We can all be forgiven. We didn't know.

Chapter 16

Fabric from Fairy Rings

Normally, I am a glass-half-full type, but I have to admit that when it comes to my faith in us humans to dig ourselves out of the messes we've created, my pessimism tends to run a bit hot. In the middle of the research for this book I spoke with a close friend, Monica Kidd, about the global plastic crisis. At the time I was feeling overwhelmed by the enormity of it and thought: *I can't write a book that offers no hope.* But I wasn't going to offer empty platitudes either.

Monica has worked on seabirds and appreciates the physical and mental necessity of having nature and beauty in the world. Monica is also a physician, a journalist, a poet, and a creator of radio and film documentaries. These apparently disparate interests have one thing in common: Monica's commitment to a just planet, where humans and wilderness are treated with empathy and respect. So while I complained that there were too many people using too many resources, her response was: we can't continue to live the way we have been, and people are the problem, but—and here's the clincher—people are also the solution.

At first, I found it hard to buy in to this argument. But as I carried on, I started to see it for myself in the research I was doing and the stories I was hearing. More and more, her words rang true and solidified my own sense of hope. Because, in direct opposition to my tendency toward pessimism, I believe that people generally want to do the right thing. We just have to know what that is.

It's hard to combat a problem of global proportions when you don't really understand the extent and complexity of it. And then, once you do, you must not be overwhelmed by it. To make meaningful headway, we must first understand the magnitude of the plastic problem and its effect on the planet and our bodies. We also have to understand that there is a viable way out that won't kill our economies while saving the planet. That plastic itself is a valuable commodity. And that just because we can't see it doesn't mean it isn't there. And finally, that combating climate change also means combating plastic production.

There are positive actions, large and small, being taken all around the world. In the late eighties I travelled by train across India. At each station, chai wallahs carried trays of tea for sale in small single-use clay cups. Once the tea was consumed, the cup was dropped, shattering into tiny pieces, reclaimed by the ground around it. I recently asked an Indian friend if the practice had ended. It had, replaced by plastic, then paper. Now the clay cups are back.

A concerned citizen in Mumbai took action at a local beach that was covered feet-deep in plastic trash. He mobilized a cleanup that took three years to complete and resulted in the return of green sea turtles, breeding there again for the first time in twenty years.

Reclaimers in South Africa and South America are cleaning up the streets and removing valuable plastic from landfills, helping the environment and making a living at the same time. In Assam, India, a couple of schoolteachers started charging students in the form of household plastic waste, to prevent them from burning it. The plastic they collected was transformed into plastic "eco-bricks" for use in construction. In Copenhagen, I saw kayaks belonging to a non-governmental organization called Greenkayak. They offer free paddling trips through the city's canals, so long as you pick up trash along the way and place it in the provided receptacles. Greenkayak operates in Denmark, Norway, Germany and Sweden. So far they have collected over forty-two thousand kilograms of plastic from the ocean. And here in my own backyard there is a local diver and self-professed former-polluter-turned-ocean-activist named Shawn Bath who founded the Clean Harbour Initiative. Shawn dives to remove fishing gear and other waste from below the surface and organizes shore beach-cleanups. To date he has pulled over sixty-eight thousand kilograms of garbage from the water.

Some grass roots beginnings have gone fullscale. The company 4Oceans started from two American surfers who were alarmed by the quantity of plastic they were encountering. They founded a business based on a 20/20 concept to fund their work: twenty dollars would pull twenty pounds of plastic trash out of the ocean. Since 2017 they have hauled over 24 million tons of plastic from the ocean. Boyan Slat went scuba diving with his family on a vacation to Greece in 2010 and was disillusioned by the plastic he saw, outnumbering the fish. He was sixteen at the time. By the age of eighteen he had founded the Ocean Cleanup, developing boom technologies to recover plastic already in the ocean, and also at the mouths of rivers, where it is intercepted before it makes it to the open ocean. His goal? To remove all marine plastic from the ocean. His company has drawn attention internationally and has become the largest ocean cleanup initiative in the world.

This morning, I turned on the radio en route to the coffee pot, my daily routine. The first thing I heard was Stella McCartney talking about using mycelium to create fabric. Mycelium are the vegetative part of fungi that grow in long strands, particularly in fairy ring fungi—mushrooms that grow in a perfect circle.

The point is, there are lots of people on the planet. And lots of them are clever and concerned and are coming up with innovative ways of dealing with plastic issues at varying scales—all having a positive impact. And just because I can't imagine them, or don't know about them, doesn't mean they don't exist. This is a huge relief.

Nature is incredibly resilient, and when we take the pressure off, it can recover. It can even make use of the waste we have left behind—the ship on Middleton, a seabird colony; the abandoned whaling stations in South Georgia and Antarctica, shelter for fur seals; even the storm-petrels, afforded protective cover by the overgrown potato garden on Baccalieu. In 2020, we saw images from around the world of wildlife entering cities during the global COVID-19 pandemic lockdown.

But the natural world has met its match with plastic. It is not something that can be adapted to, and it never goes away.

We have mostly been made aware along shorelines, where some of the plastic has come home to roost. Understandably, it has taken a long time for us to appreciate the magnitude of the problem. But we are starting to get the message loud and clear.

We should be hopeful, but not complacent. The world is looking to 2050 as the year of reckoning; if we do not get plastic waste and greenhouse gases in check by then, it may be too late. The year 2050 is not so far away.

Can countries around the world work together to find solutions? There are places to turn for inspiration. Take the COVID-19 pandemic response. In the face of a global health crisis, countries worked together, pooling resources to find solutions, and they did.

And then there is Antarctica—owned by no country but ruled by a spirit of co-operation between the signatory nations of the Antarctic Treaty. It is living proof that nations can successfully work together to study, protect and conserve.

From February 28 to March 2, 2022, representatives from over 175 member nations convened for the fifth session of the United Nations Environmental Assembly, held in Nairobi, Kenya. The meetings concluded with a historic resolution to forge a legally binding international agreement by 2024 to end plastic pollution. The agreement addresses all aspects of the plastic life cycle, from "source to sea." This landmark agreement offers solid reassurance that we are still capable of sharing in a spirit of global co-operation for the common good. And it couldn't have happened soon enough. Because it's not just about the fish we eat, the whales we are fascinated by, the beautiful sunsets, the poetry it inspires. Without the sea, we can't breathe.

———

I have spent more than a year of the last five at sea, conducting seabird surveys. Just me, my binoculars and the ocean. In serene calm and upending storms; endless horizons and pea-soup fog. I have had to ignore whale feeding frenzies to focus on overwhelming densities of seabirds that required accurate accounting. I have spent days staring across vast expanses and seen nothing at all.

You have to be prepared for nothing and also for everything. The ocean can be difficult and overwhelming, inspiring and deeply moving. I am drawn to the mystery and held by rare glimpses of the sublime. Ultimately, I have developed a deep loyalty to it.

The ocean is a well-worn metaphor, and for good reason. As if the sea was formed for the express purpose of helping us describe life's challenges and mysteries—love, wonder, , pain, death—the whole gamut. It is vast, it is deep, it is unknowable. Most of us experience the ocean near the coastline, where land and sea collide, often quite dramatically, the two elements battling it out. Cliffs standing resolute against the pounding from towering waves that have been gathering strength and resolve from many kilometres away, often building from storms that originate far out at sea. Turmoil, anger, conflict, fear.

Or the interface between land and sea can be absolutely calm, the waves arriving gently on shore—not directed by the imperative of a hurricane, but drawn gently by the pull of the moon, arriving where the ocean and shoreline have reached an understanding. The water yielding to the gentle slope of a beach, or willingly changing course to accommodate a cliff face or a complicated rocky shore. Tranquility, acceptance, comfort, peace. Throw in the sunrise, the sunset, and you have the whole arc of life

But from land our experience is mostly visual. This is where I fell in love with the ocean—from the coast. Okay, let's compare it to a love affair. We started dating on the coast. But the deep, enduring commitment? That happened far out at sea.

The open ocean has a rhythm and a tempo, and your body becomes drawn in. It can lift you and then drop you like a stone.

It can be energizing, it can be exhausting, or it can be completely tranquil. Thoughts turn to logic and pragmatism in a punchy sea, the short, staccato waves like a march.

The swell from a slow-rolling sea can lift deeply buried emotions to the surface, exposing them, unbidden, to the light. Flat, calm seas are an empty canvas, drawing on whatever emotions need to be expressed. Gratitude. Hope. Sadness. Reflection.

But these are *our* emotions. The ocean doesn't feel. It doesn't experience emotions or make judgments—it responds, impassively, to what comes its way. It simply follows the rules of nature—a change in temperature that will alter currents, increase storms, cause droughts. It will move marine nutrients, plankton and microplastics equally; it will accept the plastic barrage until there is no more room. It will spray plastic in a fine mist, back to the coast. It will swallow a whale whole, or a ship. It will provide nourishment and habitat or be poisonous. The ocean will support life, but it doesn't insist on it. The ocean doesn't make choices. We do.

Sources

APEX PREDATOR

"What is an icesheet?," National Snow and Ice Data Center, 2022, nsidc.org/learn /parts-cryosphere/ice-sheets.

United States Geological Survey, "How old is glacier ice?," USGS, accessed October 21, 2022, usgs.gov/faqs/how-old-glacier-ice?qt-news_science_products =0#qt-news_science_products.

Department of Fisheries and Oceans, "Pikialasorsuaq (North Water Polynya)," DFO, modified January 10, 2022, dfo-mpo.gc.ca/oceans/management-gestion /pikialasorsuaq-eng.html.

Ian Stirling, *Polar Bears: The Natural History of a Threatened Species* (Leaside, ON: Fitzhenry & Whiteside, 2011).

Ian Stirling and Andrew E. Derocher, "Effects of climate warming on polar bears: A review of the evidence," *Global Climate Change* 18 (2012): 2694–2706.

Michael Niaounakis, "The Problem of Marine Plastic Debris," in *Management of Marine Plastic Debris: Prevention, Recycling, and Waste Management* (Norwich, NY: William Andrew, 2017), 1–55.

CHAPTER 1. THROWAWAY LIVING

Francis K.Wiese et al., "Seabirds at risk around offshore oil platforms in the north-west Atlantic," *Marine Pollution Bulletin* 42, no. 12 (2021): 1285–90.

Science History Institute, "History and future of plastics: What are plastics and where do they come from?," *Science Matters: The Case of Plastics*, accessed October 21, 2022, sciencehistory.org/the-history-and-future-of-plastics.

Kat Eschner, "Once upon a time, exploding billiard balls were an everyday thing," *Smithsonian Magazine*, April 6, 2017, smithsonianmag.com/smart-news/once -upon-time-exploding-billiard-balls-were-everyday-thing-180962751/.

Jeffrey A. Jansen, "Plastic—it's all about molecular structure," *Plastics Engineering*, September 2016, read.nxtbook.com/wiley/plasticsengineering /september2016/consultantscorner_plastics.html.

Simon Scarr and Marco Hernandez, "Drowning in plastic: Visualising the world's addiction to plastic bottles," Reuters Graphics, September 4, 2019, graphics.reuters .com/ENVIRONMENT-PLASTIC/0100B275155/index.html.

Erik Stokstad, "World's nations start to hammer out first global treaty on plastic pollution," *Science*, February 23, 2022, science.org/content/article/world-s-nations -start-hammer-out-first-global-treaty-plastic-pollution.

Jim Valette, *The New Coal: Plastics and Climate Change* (Bennington, VT: Beyond Plastics, 2021), static1.squarespace.com/static/5eda91260bbb7e7a4bf528d8/t/616ef2 9221985319611a64e0/1634661022294/REPORT_The_New-Coal_Plastics_and _Climate-Change_10-21-2021.pdf.

Carolyn Wilke, "Plastics are showing up in the world's most remote places, including Mount Everest," *Science News*, November 20, 2020, sciencenews.org /article/plastics-remote-places-microplastics-earth-mount-everest#:~:text =Tiny%20bits%20of%20plastic%20have,by%20climbers'%20equipment %20and%20clothes.

"Circular Economy," European Parliamentary Research Service, December 2018, europarl.europa.eu/thinktank/infographics/circulareconomy/public/index.html.

European Parliament, "Plastic oceans: MEPs back EU ban on throwaway plastics by 2021," press release, October 28, 2018, europarl.europa.eu/news/en/press -room/20181018IPR16524/plastic-oceans-meps-back-eu-ban-on-throwaway -plastics-by-2021.

Roland Geyer et al., "Production, use and fate of all plastics ever made," *Science Advances* 3 (2017).

Roger Munday, "Treaty won't stem the tide of plastic waste," *Guardian*, March 11, 2022,

theguardian.com/environment/2022/mar/11/treaty-wont-stem-the-tide-of -plastic-waste.

CHAPTER 2. WILDERNESS, REWRITTEN

Jake Abrahamson, "The man who survived a polar bear attack," *Sierra*, December 15, 2014, sierraclub.org/sierra/2015-1-january-february/feature /man-who-survived-polar-bear-attack.

World Meteorological Organization, *State of the Global Climate* (Geneva: World Meteorological Organization, 2022), library.wmo.int/doc_num.php ?explnum_id=11178.

Max Liboiron, "Plastics in the gut," *Orion*, November 19, 2020, orionmagazine .org/article/plastics-in-the-gut/.

CHAPTER 3. THE HEART OF THE LABRADOR CURRENT

Brian T. Hill, "Ship collisions with iceberg database, report to PERD: Trends and analysis," NRC Technical Report TR-2005-17, 2005.

"Blue whale," NOAA Fisheries, July 21, 2022, fisheries.noaa.gov/species/blue-whale #:~:text=Blue%20whales%20are%20the%20largest,tons%20of%20krill%20a%20day.

Matthew S. Savoca et al., "Baleen whale prey consumption based on high-resolution foraging measurements," *Nature* 599 (2021): 85–90.

Carrie Arnold, "Whales eat three times more than previously thought," *National Geographic*, November 3, 2021, nationalgeographic.com/animals/article/whales -eat-three-times-more-than-thought?loggedin=true.

Lawrence Jackson et al., *The Science of Capelin: A Variable Resource* (St. John's, NL: Department of Fisheries and Oceans, 1991), waves-vagues.dfo-mpo.gc.ca /library-bibliotheque/17629.pdf.

Keith P. Lewis et al., "Forecasting capelin *Mallotus villosus* biomass in the Newfoundland shelf," *Marine Ecological Progress Series* 616 (2019): 171–83.

Douglas Hunter, "John Cabot," *The Canadian Encyclopedia*, January 7, 2008 /May 19, 2017, thecanadianencyclopedia.ca/en/article/john-cabot

William A. Montevecchi and Leslie M. Tuck, *Newfoundland Birds: Exploitation, Study, Conservation* (Cambridge, MA: Nuttal Ornithological Club, 1987).

Charles W. Townsend, ed., *Captain Cartwright and His Labrador Journal* (Boston: Dana Estes & Company, 1911), 318–19.

CHAPTER 4. QUIET AT WITLESS BAY

Chris O'Neill-Yates, Garrett Barry and Reg Sherren, "Far from a temporary move: N.L.'s cod moratorium is 25 years old," CBC News, July 2, 2017, cbc.ca /news/canada/newfoundland-labrador/cod-moratorium-twenty-five-1.4187322.

Shannon Conway, "Cod, culture, and loss: Thirty years of the cod moratorium in Newfoundland," *Active History*, July 4, 2022, activehistory.ca/2022/07/cod-culture -and-loss-thirty-years-of-the-cod-moratorium-in-newfoundland.

April Hedd et al., "Diets and distributions of Leach's storm-petrels (*Oceanodroma leucorhoa*) before and after an ecosytem shift in the Northwest Atlantic," *Canadian Journal of Zoology* 87, no. 9 (2009): 787–801.

COSEWIC, *COSEWIC Assessment and Status Report on the Leach's Storm-Petrel (Atlantic Population) Oceanodroma leucorhoa in Canada* (Ottawa: Committee on the Status of Endangered Wildlife in Canada, 2020), canada.ca/en/environment -climate-change/services/species-risk-public-registry.html.

Alexander L. Bond and Jennifer Lavers, "Effectiveness of emetics to study plastic ingestion by Leach's storm-petrels (*Oceanodroma leucohoa*)," *Marine Pollution Bulletin* 70, nos. 1–2 (2013): 171–75.

CHAPTER 5. THE FEISTINESS OF THE NORTHERN GANNET

Parks and Natural Areas Division, Department of Environment and Conservation, *Baccalieu Island Ecological Reserve Management Plan*, December 1995, gov.nl.ca /ecc/files/natural-areas-pdf-baccalieu-island-ecological-reserve.pdf.

DFO Fisheries, "Groundfish (NAFO) Division 3Ps Integrated Fisheries Management Plan," June 14, 2016, dfo-mpo.gc.ca/fisheries-peches/ifmp-gmp/groundfish-poisson -fond/groundfish-poisson-fond-div3p-2016-eng.html#toc1.1.

Thomas B. Mowbray, "Northern gannet (*Morus bassanus*)," in *Birds of the World*, ed. S.M. Billerman (Ithaca, NY: Cornell Lab of Ornithology, March 4, 2020), birdsoftheworld.org/bow/species/norgan/cur/foodhabits#feeding.

David Fifield et al., "Migratory tactics and wintering areas of northern gannets (*Morus bassanus*) breeding in North America," *Ornithological Monographs* 79 (2014): 1–63.

Warren Cornwall, "Ocean heat waves like the Pacific's deadly 'Blob' could become the new normal," *Science*, January 31, 2019, science.org/content/article/ocean-heat -waves-pacific-s-deadly-blob-could-become-new-normal.

John F. Piatt et al., "Extreme mortality and reproductive failure of common murres resulting from the northeast Pacific marine heatwave of 2014–2016," *Plos One*, January 15, 2020, doi.org/10.1371/journal.pone.0226087.

Emma Bryce, "Dead in their nests or washed ashore: Why thousands of seabirds are dying en masse," *Guardian*, September 5, 2022, theguardian.com/global- development/2022/sep/05/dead-in-their-nests-or-washed-ashore-why-thousands -of-seabirds-are-dying-en-masse.

Maria P. Dias et al., "Threats to seabirds: A global assessment," *Biological Conservation* 237 (2019): 525–37.

"Climate change indicators: Sea surface temperature," NOAA, last modified August 1, 2022, epa.gov/climate-indicators/climate-change-indicators-sea-surface -temperature#%20.

William A. Montevecchi et al., "Ocean heatwave induces breeding failure at the southern breeding limit of the northern gannet (*Morus bassanus*)," *Marine Ornithology* 9 (2021): 71–78.

CHAPTER 6. GHOST HARVEST: THE THREAT OF MACROPLASTICS

Alejandra Borunda, "This pregnant whale died with 50 pounds of plastic in her stomach:

The Mediterranean Sea is choked with plastic waste, and the sperm whale may be the latest casualty of the pollution problem," *National Geographic*, April 2, 2019, nationalgeographic.com/environment/article/dead-pregnant-whale-plastic-italy#:~:text=A%20pregnant%20sperm%20whale%20washed%20up%2C%20 dead%2C%20on%20a%20sandy,waste%20jammed%20into%20her%20belly.

Alejandra Borunda, "This young whale died with 88 pounds of plastic in its stomach:

The animal in the Philippines likely starved because its stomach was full of plastic, not food," *National Geographic*, March 18, 2019,

nationalgeographic.com/environment/2019/03/whale-dies-88-pounds-plastic -philippines/#close.

"On World Oceans Day, UNESCO reinforces the importance of preserving the largest ecosystem on the planet," UNESCO, June 8, 2020/April 21, 2022, unesco.org /en/articles/world-oceans-day-unesco-reinforces-importance-preserving-largest -ecosystem-planet.

Ian Hutton et al., "Plastic ingestion by flesh-footed (*Puffinus carneipes*) and wedge-tailed (*P. pacificus*) shearwaters," *Papers and Proceedings of the Royal Society of Tasmania* 142, no. 1 (2008): 1–6.

Barry Yeoman, "A plague of plastics: From the Arctic to Antarctica, ocean debris is killing marine wildlife—but we still have the power to stop plastic pollution,"

National Wildlife Federation, June 1, 2019, nwf.org/Home/Magazines/National
-Wildlife/2019/June-July/Conservation/Ocean-Plastic.

"New research highlights threat of marine plastics to seabirds," Circular Ocean,
accessed October 21, 2022, circularocean.eu/circularnews/new-research-seabirds
-marine-plastic-debris/.

Chris Wilcox et al., "Threat of plastic pollution to seabirds is global, pervasive,
and increasing," *Proceedings of the National Academy of Science* 112, no. 38 (2015):
11899–904.

Chris Wilcox et al., "Assessing multiple threats to seabird populations using flesh-
footed shearwaters *Ardenna carneipes* on Lord Howe Island, Australia as case
study," *Scientific Reports* 11, 7196 (2021), doi.org/10.1038/s41598-021-86702-4.

John Cooper, "Flesh-footed shearwaters deposit some 690 000 pieces of plastic
annually on Lord Howe Island," *Agreement on the Conservation of Albatrosses
and Petrels*, March 5, 2020, acap.aq/fr/actualites/dernieres-nouvelles/3962-flesh
-footed-shearwaters-deposit-some-690-000-pieces-of-plastic-annually-on
-lord-howe-island.

Jennifer Provencher et al., "Ingested plastic in a diving seabird, the thick-billed
murre (*Uria lomvia*), in the eastern Canadian Arctic," *Marine Pollution Bulletin* 60
(2010): 1406–11.

Sarah E. Nelms et al., "Plastic and marine turtles: A review and a call for research,"
ICES Journal of Marine Science 73, no. 2 (2016): 165–81.

"Albatrosses in decline from fishing and environmental change," British
Antarctic Survey, press release, November 20, 2017, bas.ac.uk/media-post
/albatrosses-in-decline-from-fishing-and-environmental-change/#:~:text=The
%20populations%20of%20wandering%2C%20black,the%20National%20Academy
%20of%20Sciences.

Library of Parliament, "Ghost fishing gear: A major source of marine plastic
pollution," *Hill Notes*, January 30, 2020, hillnotes.ca/2020/01/30/ghost-fishing
-gear-a-major-source-of-marine-plastic-pollution/#:~:text=Ghost%20fishing
%20gear%20is%20estimated,significant%20impacts%20on%20marine%20life.

Claudia Giacovelli et al., *Single-use Plastics: A Roadmap for Sustainability* (Nairobi: United Nations Environmental Programme, 2018).

"2017–2022 North Atlantic right whale unusual mortality event," NOAA Fisheries, August 30, 2022, fisheries.noaa.gov/national/marine-life-distress/2017-2022-north -atlantic-right-whale-unusual-mortality-event.

Amy R. Knowlton et al., "Effects of fishing rope strength on the severity of large whale entanglements," *Conservation Biology* 30, no. 2 (2016): 318–28.

Sarah Smellie, "New ropeless fishing technology, which can help save whales, tested off Newfoundland," CBC News, August 3, 2022, cbc.ca/news/canada/newfoundland -labrador/ropeless-fishing-technology-1.6539571.

"Why is the vaquita endangered?," Porpoise Conservation Society, 2018, porpoise.org/knowledge-base/why-is-the-vaquita-endangered/.

Stuart Wolpert, "Only 10 vaquita porpoises survive, but species may not be doomed, scientists say," *Science Daily*, May 5, 2022, sciencedaily.com/releases/2022 /05/220505143218.htm#:~:text=%22Only%2010%20vaquita%20porpoises%20survive ,ScienceDaily%2C%205%20May%202022.

"North Atlantic right whale," NOAA Fisheries, August 9, 2022, fisheries.noaa.gov /species/north-atlantic-right-whale.

"Snow Cone watch: Updates on entangled right whale mother and newborn calf," NOAA Fisheries, January 31, 2022/July 13, 2022, fisheries.noaa.gov/feature-story /snow-cone-watch-updates-entangled-right-whale-mother-and-newborn-calf.

Michelle Collins, "Update on Snow Cone—critically endangered right whale who gave birth despite chronic entanglement," Whale and Dolphin Conservation (WDC), July 27, 2022, us.whales.org/2022/07/27/update-on-snow-cone-critically -endangered-right-whale-who-gave-birth-despite-chronic-entanglement/.

V.G. Koutitonsky and G.L. Bugden, "The physical oceanography of the Gulf of Saint Lawrence: A review with emphasis on the synoptic variability of the motion," *Canadian Special Publication of Fisheries and Aquatic Sciences* 113 (1991): 57–90.

CHAPTER 8. KITTIWAKES AND THE JAPAN CURRENT

Rebecca Lindsey and Michon Scott, "What are phytoplankton?," NASA Earth Observatory, July 13, 2010, earthobservatory.nasa.gov/features/Phytoplankton.

"How much oxygen comes from the ocean? At least half of Earth's oxygen comes from the ocean," NOAA National Ocean Service, February 26, 2021, oceanservice.noaa.gov/facts/ocean-oxygen.html.

D. d'A. Laffoley et al., *The Protection and Management of the Sargasso Sea: The Golden Floating Rainforest of the Atlantic Ocean. Summary Science and Supporting Evidence Case* (Washington: Sargasso Sea Alliance, 2011), sargassoseacommission.org/sargasso-sea/threats-to-the-sargasso-sea.

Michelle L'Heureux, "What is the El Niño–Southern Oscillation (ENSO) in a nutshell?," NOAA Climate Prediction Center, May 5, 2015, climate.gov/news-features/blogs/enso/what-el-ni%C3%B10%E2%80%93southern-oscillation-enso-nutshell.

Michelle L'Heureux, "December 2021 La Niña update: visual aids", NOAA Climate Prediction Center, December 9, 2021, https://www.climate.gov/news-features/blogs/enso/december-2021-la-ni%C3%B1a-update-visual-aids

"El Niño Southern Oscillation (ENSO) Diagnostic Discussion", NOAA Climate Prediction Center, November 10, 2022, cpc.ncep.noaa.gov/products/analysis_monitoring/enso_advisory/ensodisc.shtml#:~:text=Overall%2C%20the%20coupled%20ocean%2Datmosphere,2023%20%5BFig.%206%5D.

Damond Benningfield, "The Kuroshio Current," *Science and the Sea*, January 3, 2016, scienceandthesea.org/program/201601/kuroshio-current.

Annie Feidt, "The kittiwake: Winging it, survival-wise," NPR Research News, July 23, 2011, npr.org/2011/07/23/138574244/the-kittiwake-winging-it-survival-wise.

M. Johansen et al., *International Black-Legged Kittiwake Conservation Strategy and Action Plan* (Akureyri, Iceland: Conservation of Arctic Flora and Fauna [CAFF], 2020), caff.is/cbird-caff-publications/526-international-black-legged-kittiwake-conservation-strategy-and-action-plan.

T. Nagai, "The Kuroshio current: Artery of life," *Eos*, August 27, 2019, eos.org /editors-vox/the-kuroshio-current-artery-of-life.

CHAPTER 9. FOLLOWING CURRENTS TO THE ENDS OF THE EARTH

Bethan Davies, "Increasing Antarctic sea ice: Characterstics of Antarctic sea ice," *Antarctic Glaciers*, February 26, 2021, antarcticglaciers.org/glaciers-and-climate /changing-antarctica/antarctic-sea-ice/.

Michon Scott, "Understanding climate: Antarctic sea ice extent," NOAA Climate, April 28, 2020, climate.gov/news-features/understanding-climate/understanding -climate-antarctic-sea-ice-extent.

José M. Riascos et al., "Breaking out of the comfort zone: El Niño–Southern Oscillation as a driver of trophic flows in a benthic consumer of the Humboldt Current ecosystem," *Proceedings of the Royal Society B: Biological Sciences* 284 (2017): 1–10.

"Peru's Paracas Peninsula and the Ballestas Islands: A stronghold for Humboldt seabirds," Peru Aves, accessed October 21, 2022, peruaves.org/perus-paracas -peninsula-ballestas-islands-stronghold-humboldt-seabirds/.

Nell Durfee, "Holy crap! A trip to the world's largest guano-producing islands," *Audubon*, April 27, 2018, audubon.org/news/holy-crap-trip-worlds-largest-guano -producing-islands.

Peru: Strengthening Sustainable Management of the Guano Islands, Isles, and Capes National Reserve System Project (English) (Washington: World Bank Group, 2013), documents.worldbank.org/curated/en/959321468293398621/Peru-Strengthening -Sustainable-Management-of-the-Guano-Islands-Isles-and-Capes-National -Reserve-System-Project.

Seferino Yesquén, "Peru's oil and gas overview" (PowerPoint presentation), Perú Petro, 2020, perupetro.com.pe/wps/wcm/connect/corporativo/e323fa95-24b6-463a -905e-d3cb29af2d79/FINAL+Perupetro+Presentation+Feb+6+2020.pdf?MOD =AJPERES.

L. Weilgart, "A review of the impacts of seismic airgun surveys on marine life," CBD Expert Workshop on Underwater Noise and Its Impacts on Marine and Coastal Biodiversity, February 25–27, 2014, London, UK. Available at: cbd.int/doc/?meeting=MCBEM-2014-01.

Stephen Messenger, "3,000 dolphins found dead on the coast of Peru," *Treehugger*, April 22, 2019, treehugger.com/dolphins-found-dead-coast-peru-4855034.

CHAPTER 10: THE RAVEN'S PARACHUTE: PLASTIC AND THE GULF STREAM

Mary-Louise Timmermans and John Marshall, "Understanding Arctic Ocean circulation: A review of ocean dynamics in a changing climate," *Journal of Geophysical Research: Oceans* 125 (2020), agupubs.onlinelibrary.wiley.com/doi/epdf/10.1029/2018JC014378.

"What is the cryosphere?," National Snow and Ice Data Center, 2022, nsidc.org/learn/what-cryosphere.

Michon Scott, "Sea ice withers while phytoplankton blooms in the Arctic," NOAA Climate, December 4, 2020, climate.gov/news-features/featured-images/sea-ice-withers-while-phytoplankton-blooms-arctic.

Rita Horner et al., "Ecology of sea ice biota," *Polar Biology* 12 (1992): 417–27, escholarship.org/content/qt7p59r4rx/qt7p59r4rx.pdf.

Gloria Dickie, "On this ice: Disappearing zooplankton could collapse Arctic food chain," *Arctic Deeply*, March 29, 2017, deeply.thenewhumanitarian.org/arctic/articles/2017/03/29/on-thin-ice-disappearing-zooplankton-could-collapse-arctic-food-chain

Research Council of Norway, "Zooplankton main fare for Arctic cod, marine birds and bowhead whales," *Science Daily*, June 6, 2011, sciencedaily.com/releases/2011/06/110606112526.htm.

CHAPTER 11. THE PLASTISPHERE: MICROPLASTICS, THE OCEANS
AND HUMAN HEALTH

Microplastics

Carolyn Wilke, "Plastics are showing up in the world's most remote places, including Mount Everest," Science News, November 20, 2020, sciencenews.org /article/plastics-remote-places-microplastics-earth-mount-everest#:~:text =Tiny%20bits%20of%20plastic%20have,by%20climbers'%20equipment%20 and%20clothes.

Matt Simon, "The Arctic Ocean is teeming with microfibers from clothes," *Wired*, Janurary 12, 2021, wired.com/story/the-arctic-ocean-is-teeming-with-microfibers -from-clothes/.

Mats B.O. Huserbråten et al., "Trans-polar drift-pathways of riverine European microplastic," *Scientific Report* 12 (2016): 2022, nature.com/articles/s41598-022 -07080-z.

From Pollution to Solution: A Global Assessment of Marine Litter and Plastic Pollution (Nairobi: United Nations Environmental Programme, 2012), malaysia. un.org/en/171922-pollution-solution-global-assessment-marine-litter-and-plastic -pollution.

Matthew Cole et al., "The impact of polystyrene microplastics on feeding, function and fecundity in the marine copepod *Calanus helgolandicus*," *Environmental Science and Technology* 49, no. 2 (2015): 1130–37.

Matthew Cole et al., "Effects of nylon microplastic on feeding, lipid accumulation, and moulting in a coldwater copepod," *Environmental Science and Technology* 53, no. 12 (2019): 7075–82.

Stephanie Avery-Gomm et al., "A study of wrecked dovekies (*Alle alle*) in the western North Atlantic highlights the importance of using standardized methods to quantify plastic ingestion," *Marine Pollution Bulletin* 113, nos. 1–2 (2016): 75–80.

E. Besseling et al., "Microplastic in a macro filter feeder: Humpback whale *Megaptera novaeangliae*," *Marine Pollution Bulletin* 95, no. 1 (2015): 248–52.

Human Health

"Searching for the Northwest Passage: Why did it take so long to find this sought-after trade route?," Royal Museums Greenwich, November 18, 2021, rmg.co.uk/stories/topics/search-north-west-passage.

Leanne Shapton, "Artifacts of a doomed expedition," *New York Times Magazine*, March 18, 2016, nytimes.com/interactive/2016/03/20/magazine/franklin-expedition.html.

Kala Senathirajah and Thava Palanisami, "How much microplastics are we ingesting?: Estimation of the mass of microplastics ingested," University of Newcastle (Australia), June 11, 2019, newcastle.edu.au/newsroom/featured/plastic-ingestion-by-people-could-be-equating-to-a-credit-card-a-week/how-much-microplastics-are-we-ingesting-estimation-of-the-mass-of-microplastics-ingested.

Minne Prüst et al., "The plastic brain: Neurotoxicity of micro- and nanoplastics," *Particle and Fibre Toxicology* 17, no. 4 (2020): 1–16.

R. Sung et al., "Volatile organic compounds in human milk: Methods and measurements," *Environmental Science and Technology* 41, no. 5 (2007): 1662–67.

Juliette Legler and Abraham Brouwer, "Are brominated flame retardants endocrine disruptors?," *Environment International* 29, no. 6 (2003): 879–85.

"Flame retardants," National Institute of Environmental Health Sciences, September 9, 2021, niehs.nih.gov/health/topics/agents/flame_retardants/index.cfm#:~:text=Brominated%20flame%20retardants%20%E2%80%94%20Contain%20bromine,endocrine%20disruption%20among%20other%20effects.

Joana Feiterio et al., "Health toxicity effects of brominated flame retardants: From environmental to human exposure," *Environmental Pollution* 285 (2021): 117457.

Adam Hinterthuer, "Just how harmful are bisphenol A plastics?," *Scientific American*, September 1, 2008, scientificamerican.com/article/just-how-harmful-are-bisphenol-a-plastics/.

"Bisphenol A (BPA): Use in food contact application," US Food and Drug Administration, January 2010/updated November 2014, fda.gov/food/food-additives-petitions/bisphenol-bpa-use-food-contact-application#regulations.

"EU wide bisphenol A ban expected," *Chemistry Views*, July 12, 2019, chemistryviews.org/details/news/11169386/EU_Wide_Bisphenol_A_Ban _Expected.html#:~:text=The%20use%20of%20BPA%20in,certain %20polycarbonates%20and%20epoxy%20resins.

ClientEarth, "EU court confirmed chemical BPA hazardous for health," press release, September 20, 2019,

clientearth.org/latest/press-office/press/eu-court-confirmed-chemical-bpa -hazardous-for-health/.

Hannah Zelbacker, "BPA alternatives are also harmful, researcher says," *Daily Evergreen*, September 17, 2018, dailyevergreen.com/36571/news/bpa-alternatives -are-also-harmful-researcher-says/.

Cell Press, "BPA replacements in plastics cause reproductive problems in lab mice," *ScienceDaily*, September 13, 2018, sciencedaily.com/releases/2018/09/180913113940.htm.

Phillipe D. Darbre, "Chemical components of plastics as endocrine disruptors: Overview and commentary," *Birth Defects Research* 112 (2020): 1300–1307.

Bridget Brady, "Thyroid gland: Overview," *Endocrineweb*, March 26, 2019, endocrineweb.com/conditions/thyroid-nodules/thyroid-gland-controls-bodys -metabolism-how-it-works-symptoms-hyperthyroi.

Healthwise Staff, "Thyroid hormone production and function," University of Michigan Health, Michigan Medicine, March 31, 2020, uofmhealth.org/health -library/ug1836.

Vera Koester, "Plasticizers: Benefits, trends, health, and environmental issues," *ChemistryViews*, May 5, 2015, chemistryviews.org/details/ezine/7874391/ Plasticizers__Benefits_Trends_Health_and_Environmental_Issues.html.

Niels E. Skakkebaek, "Testicular dysgenesis syndrome," *Hormone Research* 60, suppl. 3 (2003): 49.

Stephanie M. Engel et al., "Neurotoxicity of ortho-phthalates: Recommendations for critical policy reforms to protect brain development in children," *American Journal of Public Health* 111, no. 4 (2021): 687–96.

"The science behind PET," PETRA: PET Resin Association, 2015, petresin.org/science _behindpet.asp#:~:text=PET%20contains%20no%20phthalates.&text=The% 20confusion%20seems%20to%20come,made%20from%20ortho%2Dphthalic%20acid.

"Revealed: Plastic ingestion by people could be equating to a credit card a week," WWF, June 12, 2019, wwf.panda.org/wwf_news/?348337/Revealed-plastic -ingestion-by-people-could-be-equating-to-a-credit-card-a-week.

Kieran D. Cox et al., "Human consumption of microplastics," *Environmental Science and Technology* 53, no. 12 (2019): 7068–74.

Damian Carrington, "Microplastics found in human blood for first time," *Guardian*, March 24, 2022, theguardian.com/environment/2022/mar/24 /microplastics-found-in-human-blood-for-first-time.

Zehua Yan et al., "Analysis of microplastics in human feces reveals a correlation between fecal microplastics and inflammatory bowel disease status," *Environmental Science and Technology* 56, no. 1 (2022): 189–93.

Minne Prüst et al., "The plastic brain: Neurotoxicity of micro- and nanoplastics," *Particle and Fibre Toxicology* 17, no. 24 (2020), doi.org/10.1186/s12989-020-00358-y.

Phillipe D. Darbre, "Environmental contaminants: Environmental estrogens— hazard characterization," *Encyclopedia of Food Safety* 2 (2014): 323–31.

Antonia Ragusa et al., "Plasticenta: First evidence of microplastics in human placenta," *Environment International* 146 (2021), sciencedirect.com/science/article /pii/S0160412020322297.

Linda S. Birnbaum and Daniele F. Staskal, "Brominated flame retardants: Cause for concern?," *Environmental Health Perspectives* 112, no. 1 (2004): 9–17.

Kiyoshi Yamauchi, "Subchapter 102A: Tetrabromobisphenol A," in *Handbook of Hormones: Comparative Endocrinology for Basic and Clinical Research* (Cambridge, MA: Academic Press, 2016), 593–95.

Emmanuel Sunday Okeke et al., "Review of the environmental occurrence, ana- lytical techniques, degradation and toxicity of TBBPA and its derivatives," *Environmental Research* 105 (2022), doi.org/10.1016/j.envres.2021.112594.

"Facts about benzene," Centers for Disease Control and Prevention (CDC), April 4, 2018, emergency.cdc.gov/agent/benzene/basics/facts.asp#:~:text=The %20Department%20of%20Health%20and,of%20the%20blood%2Dforming %20organs.

Lisa Erdle, *Problems with polystyrene foam: environmental fate and effects in the Great Lakes* (Georgian Bay Forever: ON, 2019), georgianbayforever.org /Polystyrene/GBFReportPSFoam/2/

"Extruded polystyrene foam vs. Styrofoam: What's the difference?", Foam Equipment and Consulting Co., foamequipment.com/blog/bid/33863/what-is -styrofoam

CHAPTER 12. WHALES ON TOAST

Johnny Briggs, "Whaling and seal hunting defined South Georgia—but then crashed:

Journey to the jewel of the polar crown reveals a stark history," Pew Charitable Trusts, July 13, 2017, pewtrusts.org/en/research-and-analysis/articles/2017/07/13 /whaling-and-seal-hunting-defined-south-georgia-but-then-crashed.

Richard York, "Why petroleum did not save the whales," *Socius: Sociological Research for a Dynamic World* 3 (2017): 1–13.

"The future of petrochemicals: Towards a more sustainable chemical industry," International Energy Agency, iea.org/reports/the-future-of-petrochemicals.

Amanda Kistler and Carroll Muffett, eds., "Plastic and climate: The hidden costs of a plastic planet," Center for International Environmental Law (CIEL), accessed October 21, 2022, ciel.org/plasticandclimate/.

Hale Türkes, "Saudi Aramco to invest over $100B in petrochemicals," Anadolu Agency (AA), November 27, 2018, aa.com.tr/en/energy/news-from-companies/ saudi-aramco-to-invest-over-100b-in-petrochemicals/22473.

CHAPTER 13. OUT OF SIGHT, OUT OF MIND

Chaiwat Satyaem, "'Garbage island' will take 10 days to clear," *Bangkok Post*, February 10, 2017, bangkokpost.com/thailand/general/1196369/garbage-island -floats-back-into-view.

Teale Phelps Banderoff and Sam Cook, *Masks on the Beach: The Impact of* COVID-19 *on Marine Plastic Pollution* (Oceans Asia, 2020), oceansasia.org/covid-19-facemasks/.

David Farrier, "Hand in glove: The false promise of plastics," *Orion Magazine*, September 10, 2020, orionmagazine.org/article/hand-in-glove/.

Marc Barton, "The history of surgical gloves," *Past Medical History*, September 19, 2018, pastmedicalhistory.co.uk/the-history-of-surgical-gloves/.

Hannah Ritchie and Max Roser, "Plastic pollution," *Our World in Data*, September 2018/April 2022, ourworldindata.org/plastic-pollution.

"The mobile Mobius: A history of the recycling symbol," Green Dining Alliance (GDA), September 2, 2022, greendiningalliance.org/2016/03/the-mobile-mobius-a-history-of-the-recycling-symbol/.

Rose Secrest, "Plastic wrap," *How Plastics Are Made*, accessed October 21, 2022, madehow.com/Volume-2/Plastic-Wrap.html.

Xiangyu Jie et al., "Microwave-initiated catalytic deconstruction of plastic waste into hydrogen and high-value carbons," *Nature Catalysis* 3 (2020): 902–12.

Will Kenton, "Greenwashing," *Investopedia*, March 22, 2022, investopedia.com/terms/g/greenwashing.asp#:~:text=Greenwashing%20is%20the%20process%20of,company's%20products%20are%20environmentally%20friendly.

David Rachelson, "What is recycling contamination, and why does it matter?," *Rubicon*, December 4, 2017, rubicon.com/blog/recycling-contamination/.

From Pollution to Solution: A Global Assessment of Marine Litter and Plastic Pollution (Nairobi: United Nations Environmental Programme, 2012), malaysia.un.org/en /171922-pollution-solution-global-assessment-marine-litter-and-plastic-pollution.

Imogen E. Napper and Richard C. Thompson, "Environmental deterioration of biodegradable, oxo-biodegradable, compostable, and conventional plastic carrier bags in the sea, soil, and open-air over a 3-year period," *Environmental Science and Technology* 53, no. 9 (2019): 4775–83.

Irena Slav, "How much crude oil does plastic production really consume?," *Oil Price*, October 10, 2019, oilprice.com/Energy/Energy-General/How-Much-Crude-Oil-Does-Plastic-Production-Really-Consume.html.

"How much of ocean plastics comes from land and marine sources?," *Our World in Data*, September 2018/April 2022, ourworldindata.org/plastic-pollution#how-much-of-ocean-plastics-come-from-land-and-marine-sources.

Rozanna Latiff, "Malaysia to send 3,000 tonnes of plastic waste back to countries of origin," Reuters, May 28, 2019, reuters.com/article/us-malaysia-waste-idUSKCN1SY0M7.

Roland Geyer et al., "Production, use, and fate of all plastics ever made," *Science Advances* 3, no. 7 (July 19, 2017), science.org/doi/10.1126/sciadv.1700782.

Shosuki Yoshida et al., "A bacterium that degrades and assimilates poly(ethylene terephthalate)," *Science* 361, no. 6278 (2016): 1196–99, science.org/doi/10.1126/science.aad6359.

Brandon C. Knott et al., "Characterization and engineering of a two-enzyme system for plastics depolymerization," *Proceedings of the National Academy of Science (PNAS)* 117, no. 41 (2020): 25476–85, pnas.org/doi/10.1073/pnas.2006753117.

Till Tiso et al., "Towards bio-upcycling of polyethylene terephthalate," *Metabolic Engineering* 66 (2021): 167–78, sciencedirect.com/science/article/pii/S1096717621000471.

CHAPTER 14. REFRAMING PLASTIC: THE POTENTIAL OF A
CIRCULAR ECONOMY

"Designing out plastic pollution," Ellen MacArther Foundation, accessed October 21, 2022, ellenmacarthurfoundation.org/our-work/activities/new-plastics-economy#:~:text=The%20Global%20Commitment&text=In

%20October%202018%2C%20in%20collaboration,at%20its%20source%2C
%20by%202025.

"The circular economy in detail," Ellen MacArthur Foundation, 2017,
ellenmacarthurfoundation.org/explore/the-circular-economy-in-detail?gclid
=CjwKCAiAl4WABhAJEiwATUnEF2100Vxjuo6ZZlNCo_pUMaeYde
93QboyYdBFycF81ZpmPS-ohoCJbRoCPJsQAvD_BwE.

"Strategy on Zero Plastic Waste," Canadian Council of Ministers of the Environment
(CCME), November 23, 2018, ccme.ca/en/res/strategyonzeroplasticwaste.pdf

"Canada-wide Action Plan on Zero Plastic Waste: Phase 1," Canadian Council of
Ministers of the Environment (CCME), 2019, ccme.ca/en/res/1589_ccmecanada
-wideactionplanonzeroplasticwaste_en-secured.pdf.

"Canada-wide Action Plan on Zero Plastic Waste: Phase 2," Canadian Council of
Ministers of the Environment (CCME), 2020, ccme.ca/en/res/ccmephase2actionplan
_en-external-secured.pdf.

Ian Tiseo, "Market value of plastic recycling worldwide in 2019 and 2027 (in billion
U.S. dollars)," *Statista*, January 26, 2021, statista.com/statistics/987522/global-market
-size-plastic-recycling/.

"European Commission adopts Circular Economy Action Plan," International
Institute for Sustainable Development (IISD), May 28, 2020, sdg.iisd.org/news
/european-commission-adopts-circular-economy-action-plan/#:~:text=The
%20plan%20aims%20to%20reduce,EU's%20carbon%20and%20material%20
footprint.&text=It%20introduces%20legislative%20and%20non,EU%20level
%20brings%20added%20value.

Elisabeth Braw, "Five countries moving ahead of the pack on circular economy
legislation," *Guardian*, October 29, 2014, theguardian.com/sustainable
-business/2014/oct/29/countries-eu-circular-economy-legislation-denmark
-sweden-scotland.

"Parties to the Basel Convention on the Control of Transboundary Movements
of Hazardous Wastes and Their Disposal," Basel Convention, 2011, basel.int
/?tabid=4499.

"Why hasn't the United States ratified the Basel Convention?," US Environmental Protection Agency (EPA), February 7, 2022, epa.gov/hwgenerators/frequent -questions-international-agreements-transboundary-shipments-waste#basel.

"US set to become the world's biggest criminal trafficker in plastic waste," Basel Action Network, December 10, 2010, ban.org/news/2020/12/10/us-set-to-become -the-worlds-biggest-criminal-trafficker-in-plastic-waste.

Jacob Wallace, "Greens say U.S.–Canada plastic shipment violates treaty," Basel Action Network, January 5, 2021, ban.org/news-new/2021/1/5/greens-say-us -canada-plastic-shipping-violates-treaty.

"Ocean Plastics Charter," Government of Canada, December 9, 2021, canada.ca/en /environment-climate-change/services/managing-reducing-waste/international -commitments/ocean-plastics-charter.html.

Environment and Climate Change Canada, *Economic Study of the Canadian Plastic Industry, Markets and Waste: Summary Report to Environment and Climate Change Canada* (Gatineau, QC: Environment and Climate Change Canada, 2019), publications.gc.ca/collections/collection_2019/eccc/En4-366-1-2019-eng.pdf.

Library of Parliament, "Executive summary: Global marine plastic pollution: sources, solutions and Canada's role," *Hill Notes*, February 10, 2020, hillnotes. ca/2020/02/10/executive-summary-global-marine-plastic-pollution-sources -solutions-and-canadas-role/.

Library of Parliament, "Ghost fishing gear: A major source of marine plastic pollution," *Hill Notes*, January 30, 2020, hillnotes.ca/2020/01/30/ghost-fishing -gear-a-major-source-of-marine-plastic-pollution/#:~:text=Ghost%20fishing %20gear%20is%20estimated,significant%20impacts%20on%20marine%20life.

Canadian Council of Ministers of the Environment (CCME), "Backgrounder: A Canada-wide action plan for extented producer responsibility," Canadian Intergovenmental Conference, Secretariat, October 29, 2009, scics.ca/en/product -produit/backgrounder-a-canada-wide-action-plan-for-extended-producer -responsibility/.

Environment and Climate Change Canada, "Canada one-step closer to zero plastic waste by 2030," press release, October 7, 2020, canada.ca/en/environment-climate- change/news/2020/10/canada-one-step-closer-to-zero-plastic-waste-by-2030.html.

"Ghost Gear Fund," Fisheries and Oceans Canada, June 21, 2022, dfo-mpo.gc.ca /fisheries-peches/management-gestion/ghostgear-equipementfantome/program -programme/projects-projets-eng.html.

Microbeads in Toiletries Regulations (SOR/2017-111), Canadian Environmental Protection Act, 1999/ Registration 2017-06-02, laws-lois.justice.gc.ca/eng /regulations/SOR-2017-111/page-1.html#h-839363.

"Canada Gazette, Part I, Volume 154, Number 41: Order adding a toxic substance to Schedule 1 to the Canadian Environmental Protection Act, 1999," *Canadian Environmental Protection Act, 1999*, October 10, 2020, gazette.gc.ca/rp-pr/p1/2020 /2020-10-10/html/reg1-eng.html.

Carol E. Landisman and Barry W. Connors, "Long-term modulation of electrical synapses in the mammalian thalamus," *Science* 310, no. 5755 (December 16, 2005): 1809–13.

Morten Anderson, "Bacteria can break down plastic," DTU, September 30, 2019, dtu.dk/english/news/topics/the-big-plastic-challenge/nyhed?id=afa835a7-eb08 -4f05-a5ee-9212464dcf26.

Damien Carrington, "Scientists create mutant enzyme that recycles plastic bottles in hours:

Bacterial enzyme originally found in compost can be used to make high-quality new bottles," *Guardian*, April 8, 2020, theguardian.com/environment/2020/ apr/08/scientists-create-mutant-enzyme-that-recycles-plastic-bottles-in-hours.

Damien Carrington, "New super-enzyme eats plastic bottles six times faster: Breakthrough that builds on plastic-eating bugs first discovered by Japan in 2016 promises to enable full recycling," *Guardian*, September 28, 2020, theguardian. com/environment/2020/sep/28/new-super-enzyme-eats-plastic-bottles-six-times- faster?CMP=share_btn_link.

María José Cárdenas Espinosa et al., "Toward biorecycling: Isolation of a soil bacterium that grows on a polyurethane oligomer and monomer," *Frontiers in Microbiology* 11, no. 404 (2020): 1–9, frontiersin.org/articles/10.3389/fmicb.2020.00404/full.

Damien Carrington, "Scientists find bug that feasts on toxic plastic: Bacterium is able to break down polyurethane, which is widely used but rarely recycled,"

Guardian, March 27, 2020, theguardian.com/environment/2020/mar/27/scientists -find-bug-that-feasts-on-toxic-plastic.

Scarlett Evans, "Turning waste into power: The plastic to fuel projects," *Power Technology*, September 11, 2020, power-technology.com/comment/plastic-to -fuel/#:~:text=Plastic%20to%20diesel,petroleum%20and%20other%20fuel %20products.

"The potential economic impact of advanced recycling and recovery facilities in the United States," American Chemisty Council, April 12, 2022, americanchemistry.com /better-policy-regulation/plastics/advanced-recycling/resources/the-potential -economic-impact-of-advanced-recycling-and-recovery-facilities-in-the-united-states.

Xiangyu Jie et al., "Microwave-initiated catalytic deconstruction of plastic waste into hydrogen and high-value carbons," *Nature Catalysis* 3 (2020): 902–12.

CHAPTER 15. NEW BEGINNINGS: LAND OF ICE AND HOPE

Hadoram Shirihai, *The Complete Guide to Antarctic Wildlife: Birds and Marine Mammals of the Antarctic Continent and the Southern Ocean* (Princeton, NJ: Princeton University Press, 2008).

"Krill Matters," Australian Antarctic Program, accessed October 21, 2022, antarctica.gov.au/news/explore-antarctica/krill/.

"Crabeater seal," Australian Antarctic Program, March 20, 2018, antarctica.gov. au/about-antarctica/animals/seals/crabeater-seal/.

John Turner et al., "Record low Antarctic sea ice cover in February 2022," *Geophysical Research Letters* 49 (2022), agupubs.onlinelibrary.wiley.com/ doi/10.1029/2022GL098904.

CHAPTER 16. FABRIC FROM FAIRY RINGS

"The 4Ocean trash tracker," 4Ocean, August 21, 2022, 4ocean.com/pages/our-impact.

"We are the ocean cleanup," The Ocean Cleanup, 2022, theoceancleanup.com/about/.

UN Environmental Programme (UNEP), "UN Environment Assembly opens with all eyes on a global agreement on plastic pollution," press release, February 28, 2022, unep.org/news-and-stories/press-release/un-environment-assembly-opens-all-eyes-global-agreement-plastic.

"What you need to know about the plastic pollution resolution," UN Environmental Programme (UNEP), March 2, 2022, unep.org/news-and-stories/story/what-you-need-know-about-plastic-pollution-resolution.

Acknowledgements

This book delves into several areas of discipline. In some cases, I knew only enough to ask the right questions. The writing has been an incredible voyage of discovery in itself. For generously sharing their experience, perspective and expertise in all things biological, meteorological, chemical, oceanographical, historical and medical, I owe an enormous debt of gratitude to the following people: Alex Bond, Anne Drover, Arielle Hogan, Patricia Hunt, Ian Jones, Tom Kehoe, Monica Kidd, Jack Lawson, Bruce Mactavish, Fran Mobray, Bill Montevecchi, David Nettleship, John Piatt, John Pratt, Mark Ringuette, Rosann Seviour, Steve Snook, Kate Stafford, Ian Stirling, Bruce Whiffen and Sabina Wilhelm.

I would like to give special thanks to Brad de Young, who spent a generous amount of time discussing the complicated but fascinating field of physical oceanography with me. He offered fresh perspectives and answered all my questions thoroughly and patiently.

He tutored me, basically.

Thanks also to Adventure Canada and Hurtigruten for introducing me to cultures and creatures from around the world.

Invaluable financial assistance was provided by the Canada Council for the Arts.

Reading early drafts of this book was a slog; I know it was. For persevering and providing valuable input, I would like to thank Gail Hogan, Monica Kidd, Mary Lewis, Lisa Moore, John Pratt and Michael Crummey. Actually, Michael read every draft. And provided in-house editorial advice. And moral support. And answers to "what do you call that thing . . . what's the word for . . . starts with an *m* . . ." And every other kind of support. For all of it, Michael—my deepest love and gratitude.

To Kristin Cochrane, at the helm of Penguin Random House Canada; Martha Kanya-Forstner, publisher at Knopf Canada; and my agent Martha Webb at CookeMcDermid Agency—I want to thank you for believing that this book was a good idea, and making it happen.

Amanda Betts, my editor at Knopf Canada—saying you helped shape the book is like saying glaciers helped shape the landscape. Thank you for applying your masterful eye, sharp focus and intelligence to the book. And thank you also for being such a joy to work with. I think we make a great team.

John Pratt introduced me to the wonder of birds, John Piatt cast a seabird spell, Bruce Mactavish shared a lifetime of bird experience, and Bill Montevecchi taught me what questions to ask and how to answer them, while always working hard and celebrating the beauty—he leads by example. Carina Gjerdrum has given me the opportunity to spend long periods of time far out at sea. The personal impact has been immense. I want to thank you all for your mentorship and friendship over the years, and for leading me to the place that made writing this book feel like an urgent need.

For your great company and for sharing your knowledge of plastic in the Arctic, I am so grateful to you, Jayco Tatatuipik. Nakurmiik.

I also owe special thanks to Ian Jones—for your beautiful, informed artwork, your insights and your tremendous generosity. The book has benefited greatly from your gifts.

And to my family. Arielle, Robin, Emma, Michael and Mike. We have faced the roughest waters a family can navigate, in the only way imaginable—together. Nothing would have been possible without you. And last but not least—thank you, Ben. Moving forward into the light has been the only option. Because to do anything else would be to dishonour your love for life and your desire to bring joy to the people you loved. You are on every page, and in every breath.

HOLLY HOGAN is a writer and wildlife biologist with a focus on seabirds. During her more than thirty years as a scientist, she has spent about a thousand days at sea conducting avian and marine mammal surveys and providing educational programming with expedition teams. Her work has taken her to the Arctic and Antarctic Oceans, and every latitude in between. She has been interviewed for CBC Radio, appears in a National Film Board series called *Ocean School*, and provided expertise on seabirds and the impact of marine plastic for the award-winning documentary *Hell or Clean Water* (2021). Holly is a mother of three and lives in St. John's, Newfoundland with her husband, Michael, and an assortment of cats and dogs, depending on the day.